SPEED OF LIGHT IS INCREASED EQUALLY EVERYWHERE

New Revolutionary Theory in Physics

Dr/Mohamed Abdelwhab Husseiny

October 6 University, Edfu, 81718, Egypt
E-mail: mohamed_abdelwhab1987@yahoo.com

This book is dedicated to

*Heroes of January 25 revolution for inspir-
ing me that anything is possible.*

My parents, Abdelwhab and Nagat.

My brother, Eslam and my sisters.

CONTENTS

PREFACE

This book use easy scientific language where every term is explained thoroughly. After reading this book, you will know many secrets about nature of the universe such as the relationship between the speed of light and the observer effect, the secrets of dark energy and dark matter and you will be able to explain the double slit experiment.

I spent eight years wondering why speed of light is constant, because it is so difficult for any body or particle to still moving with a constant speed for long time, so constancy of speed of light since the birth of the universe seems impossible.

-I find that there is no cause to believe in constancy of speed of light, rather it will be more logical and acceptable if we postulate that speed of light is increased equally everywhere, so speed of light will be the same for everyone everywhere and thus the special relativity still working correctly with the new suggested postulate.

-I find that the speed of light goes to be decreased at observation regarding to the distance that the light travels as the observation process takes time and as the speed of light is increased with time while in absence of observation light has absolute speed that is is increased equally everywhere.

-I find that "Redshift" occurs as the speed of light is increased with time thus wavelength of light must be increase with time also.

-I find that the universal gravitational constant (G) is increased with time.

-I find that the dark matter doesn't exist where $(\Delta G) / r$ = constant value $= 5.11 \times 10\text{-}37$, thus the more orbit radius the more difference in (G) between (G) at the center of the orbit and (G) at the orbit .

-I find that the cause of the observer effect is:
"Each particle is exist in a single universe in absence of observation where each universe has a different space dimensions (Simultaneous Multiverse)"
"All particles exists in "Time separated universe" at observation where speed of light between bodies is relative and time is relative".

-I find that mass of bodies is increased at observation as the speed of light is decreased at observation..

-I find that momentum of bodies is increased at observation as the speed of light is decreased at observation.

-I find that wavelength is decreased at observation as the speed of light is decreased at observation.

- I find that there is a fifth force and that force is the cause behind acceleration of light, it is equal to $1.43 \times 10\text{-}48$ N.

- I find that light is accelerate with $6.9 \times 10\text{-}10$ m/s2.

And other interesting secrets.....

Special theory of relativity:
-The speed of light in vacuum has the same value in all inertial frames, regardless of the velocity of the observer or the velocity of the source emitting the light.

First probability:
-Speed of light is constant.

Second probability:
-Speed of light is changed equally in all frames of reference.

PART ONE:

Acceleration of light

CHAPTER 1:

The idea behind the change of speed of light

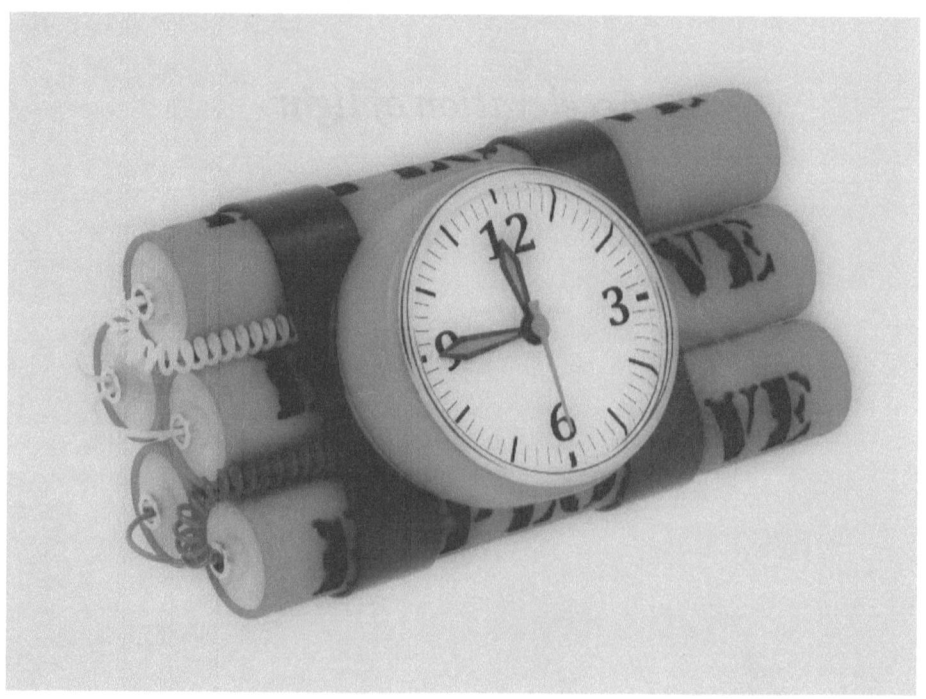

The birth of space and time of the static universe differs from that of the expanded one. In a static universe, space and time are already exist and they have positive values with no rise from zero where space is neither expanding nor contracting and time is changed (Fig. 1).

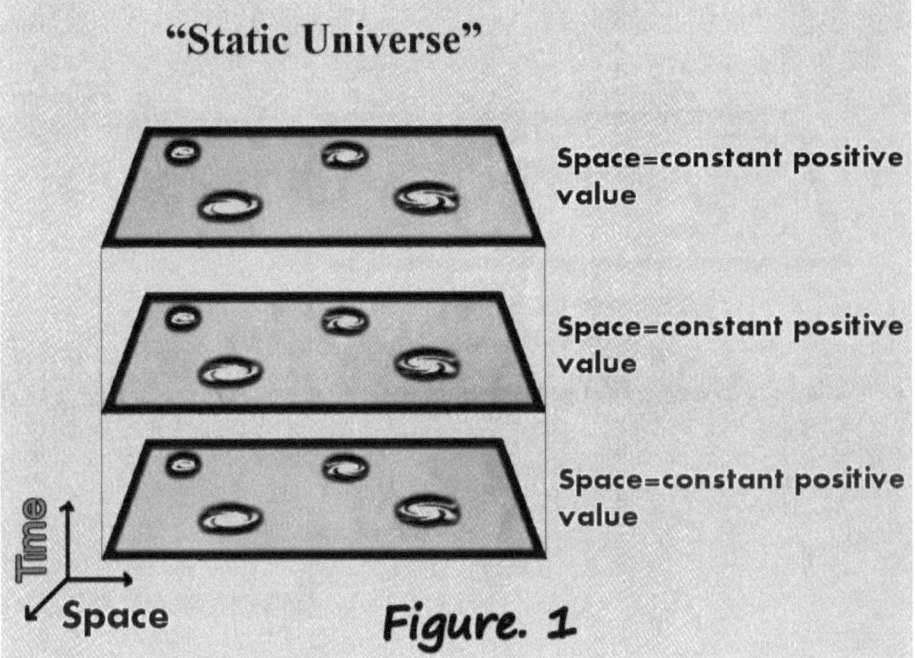

Figure. 1

While in the expanded universe, space and time are originated from zero and increased or expanded (Fig. 2).

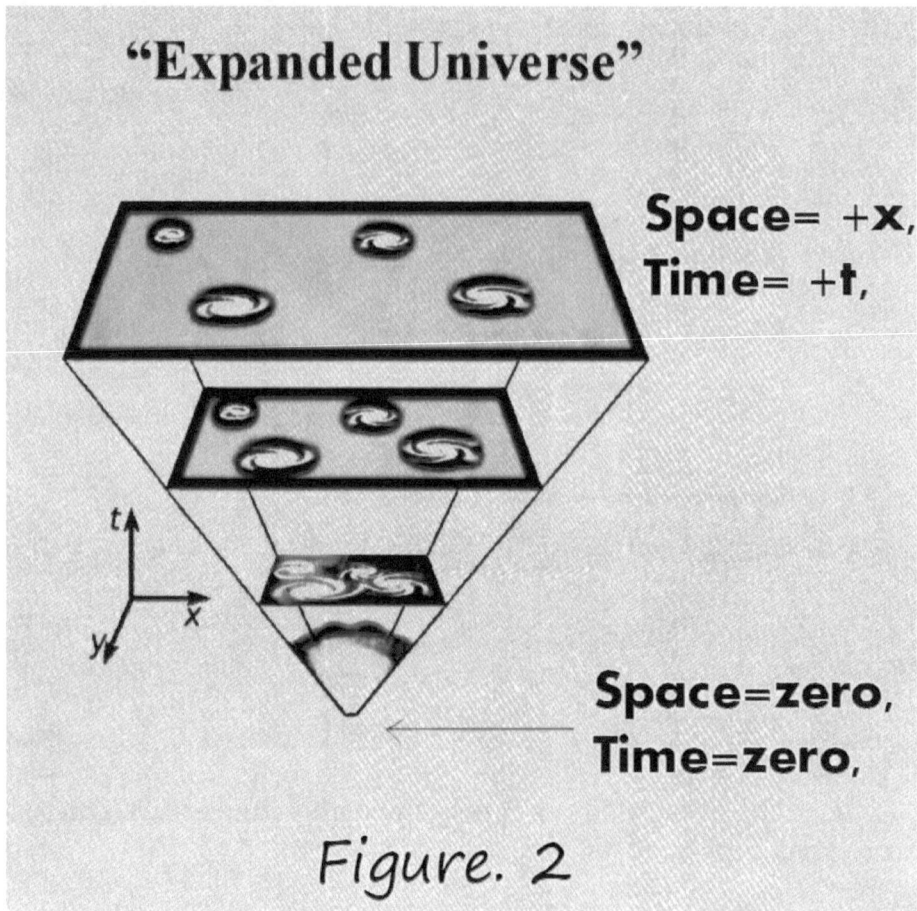

"Expanded Universe"

Space= +x,
Time= +t,

Space=zero,
Time=zero,

Figure. 2

This book suggests that if the static universe is temporary, it must have an end that at it time is no longer exists; this means the value of time at this final stage is equal to zero (Fig. 3).

"Temporary Static Universe"

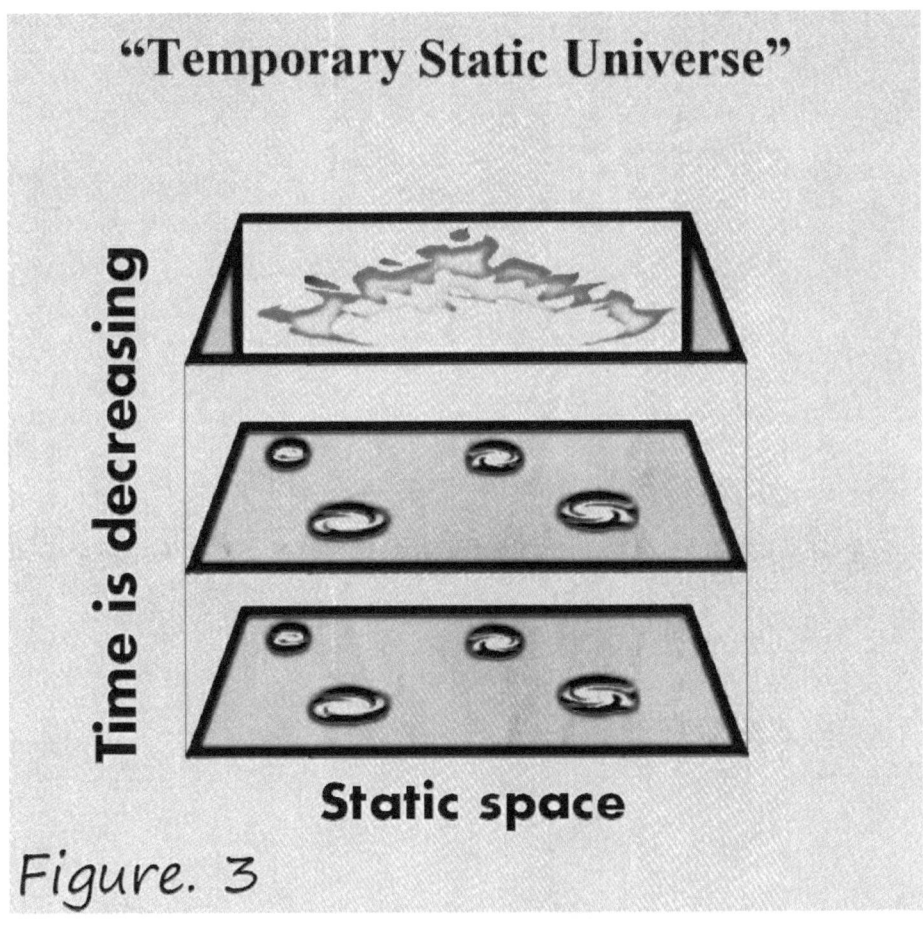

Figure. 3

The change of time from positive value to zero refers to decreasing or contraction of time of the static universe. Accordingly I considered that the static universe has a timer which is similar to the bomb's timer and that timer is the main property of the temporary static universe.

As the space in the static universe is fixed or static and time is contracting, speed of light must be increased in this cosmological model,

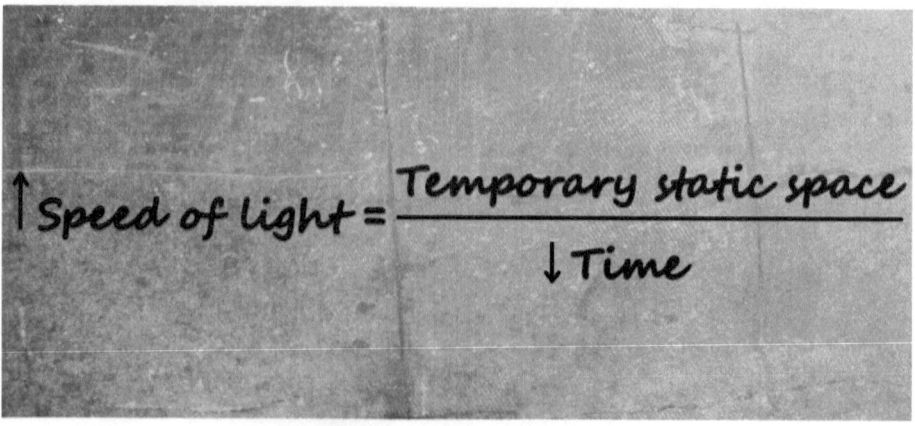

therefore I postulated the following:

"Speed of light is increased equally everywhere".

or in other words

"the change of speed of light (Δc) is absolute" **(Fig. 4,5&6).**

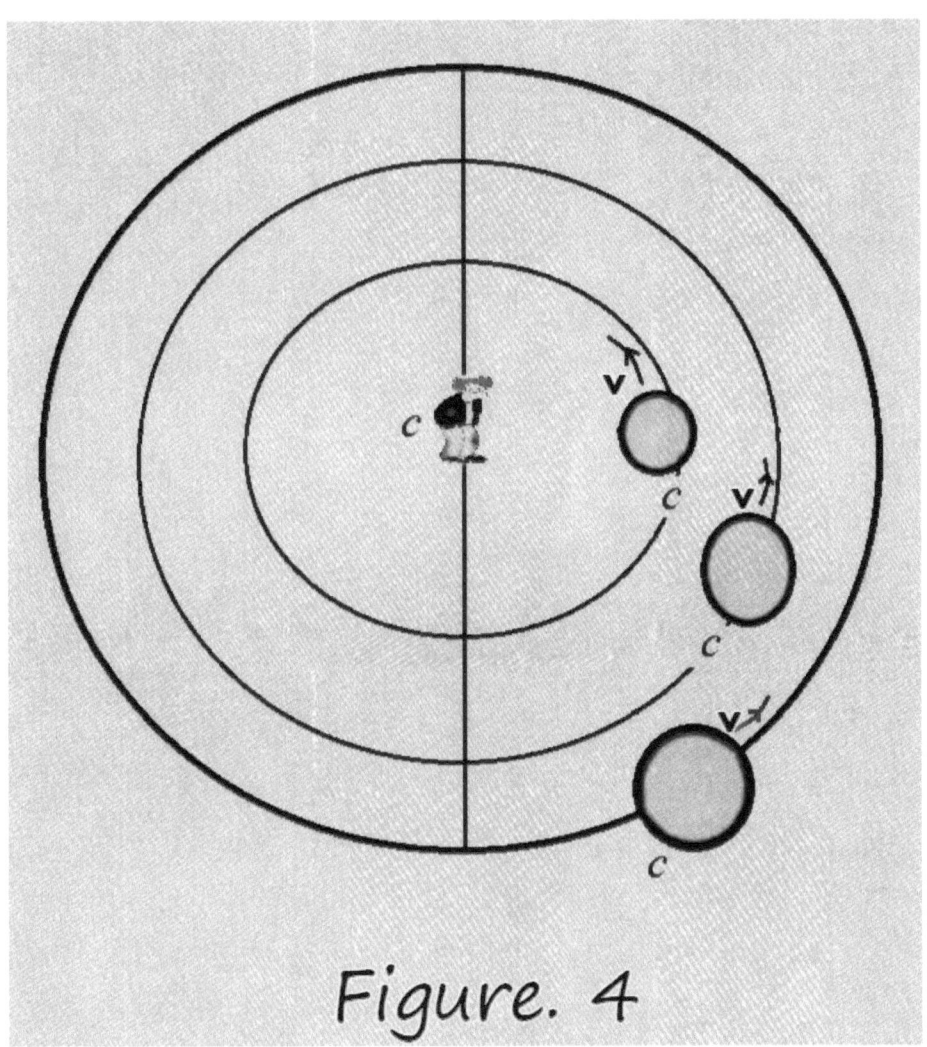

Figure. 4

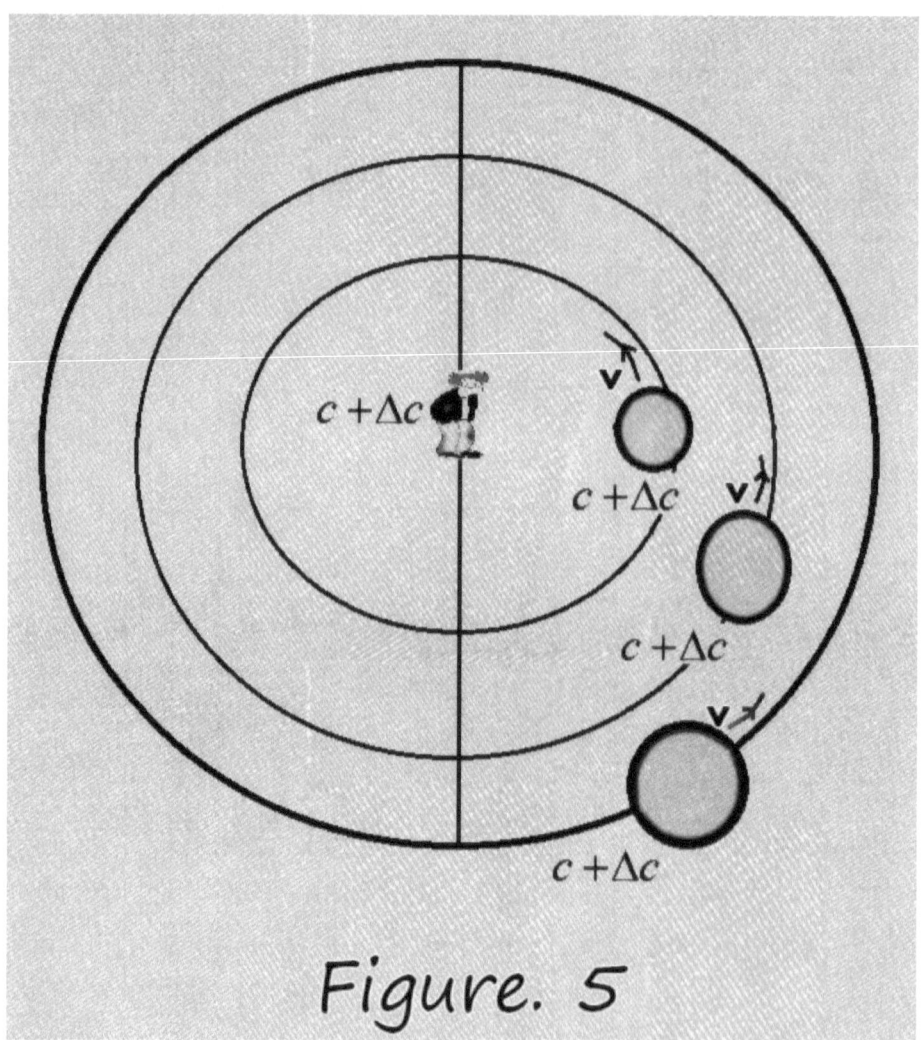

Figure. 5

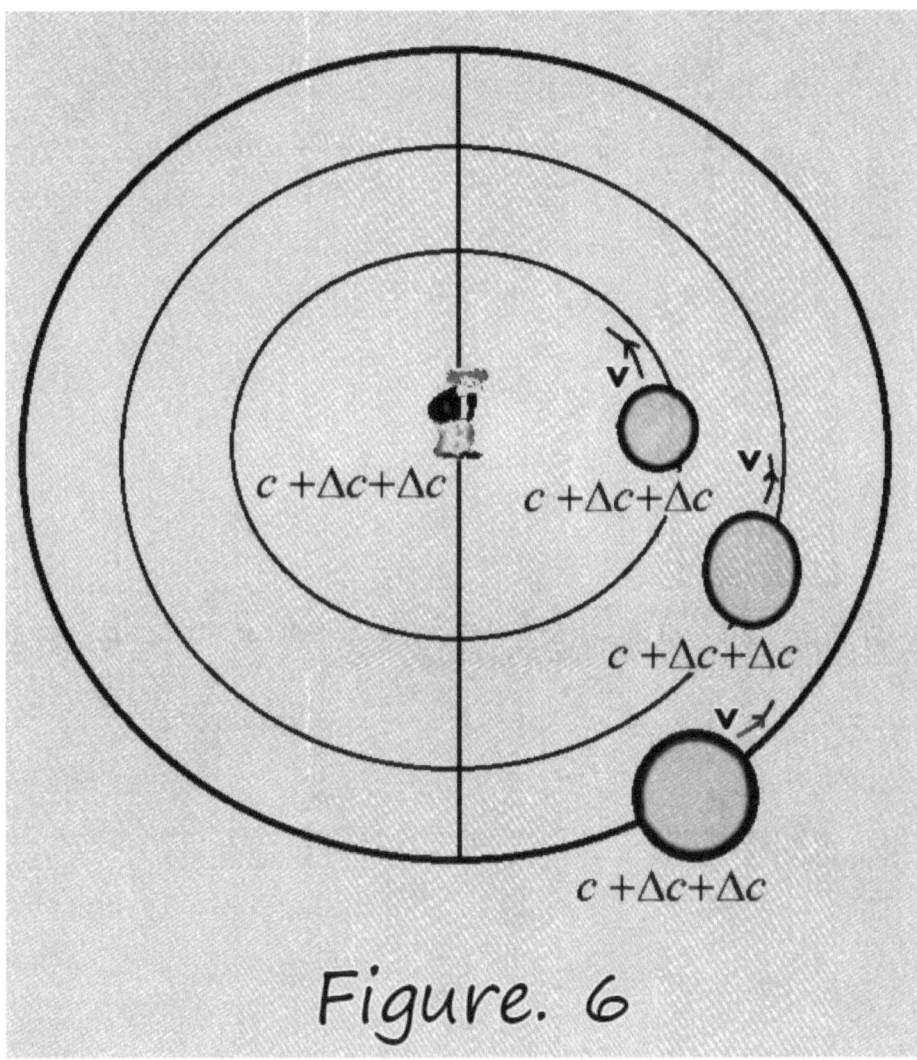

Figure. 6

The mentioned postulate shows that the speed of light is increased with time thus light is accelerating in the temporary static universe. (Fig. 7).

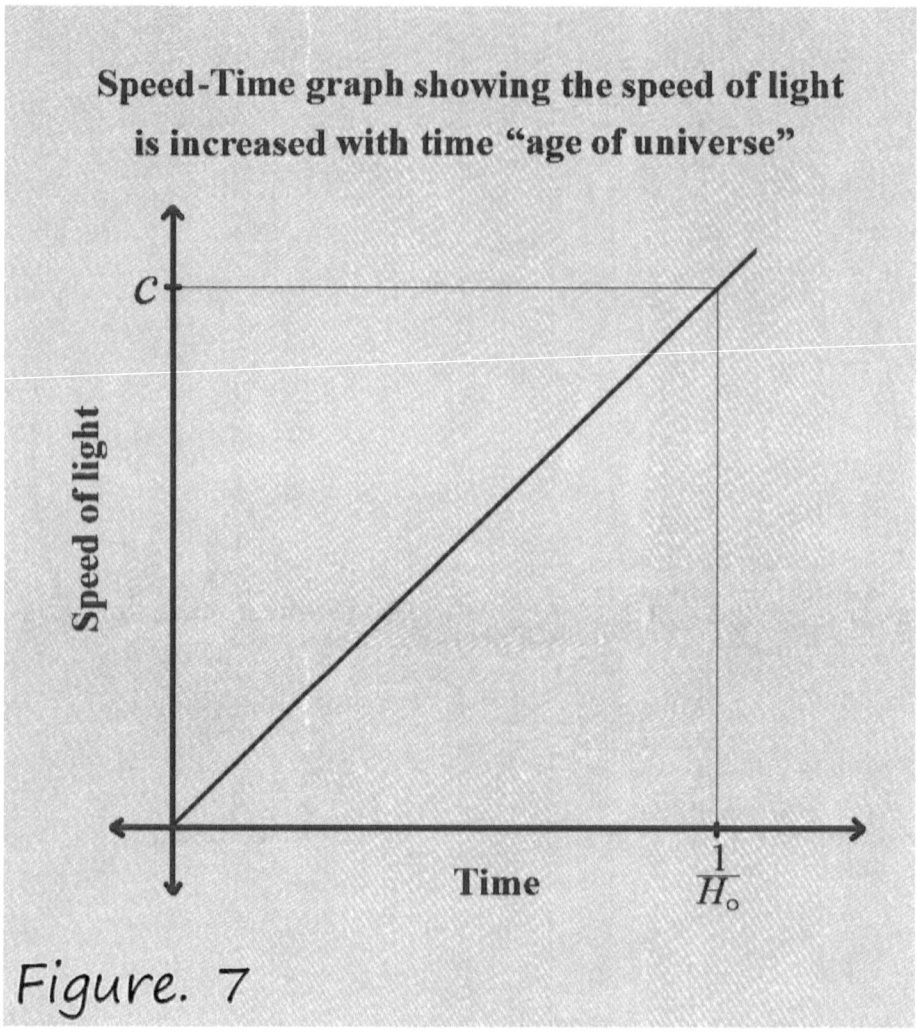

Speed-Time graph showing the speed of light is increased with time "age of universe"

Figure. 7

The new postulate agrees with the famous postulate of special relativity that states "speed of light is the same in all frames of reference" as the speed of light is increased equally in all frames regardless the motion of the frames and thus speed of light is still the same in all frames of reference.

Regarding to nature of the temporary-static universe, the new mentioned postulate is necessary to compensate the decreasing in time where space is still static.

Speed of Light is Increased Equally Everywhere

CHAPTER 2:

Relativistic acceleration of light

As the change of speed of light (Δc) is absolute in all frames of reference and time is dilated with motion (according to special relativity), I find that acceleration of light in moving frames must be relative for observers at rest (Fig.8) where the value of acceleration of light in moving frame with respect to the moving observer differs from value of acceleration of light in that frame with respect to the observer at rest or the stationary frame of reference. I find that there is a relation between acceleration of light in a moving frame and velocity of that frame with respect to the observer at rest.

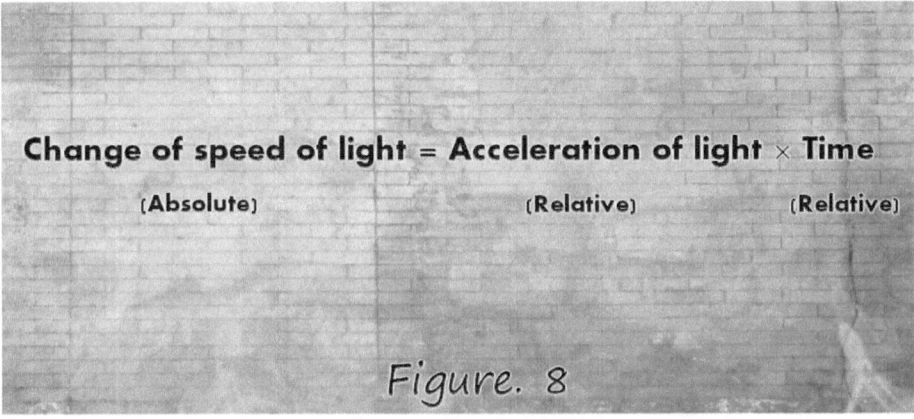

Change of speed of light = Acceleration of light × Time

(Absolute)   (Relative)   (Relative)

Figure. 8

Thus if (a_I') denotes the acceleration of light in the moving frame with respect to the moving observer and (a_I) denotes the acceleration of light in the movingframe with respect to the stationary frame of reference, the relation between a_I' & a_I can be given as follows:

$$\Delta c = \Delta c',$$

$$a_l.t = a_l'.t',$$

And as,

$$t = \gamma t',$$

I get,

$$a_l.\gamma t' = a_l'.t',$$

$$a_l.\gamma = a_l',$$

$$a_l = \frac{a_l'}{\gamma},(1)$$

where (γ) denotes the lorentz factor.
Regarding to (Eq. 1), the acceleration of light in inertial frames is decreased with respect to the stationary frames of reference (Fig. 9-13).

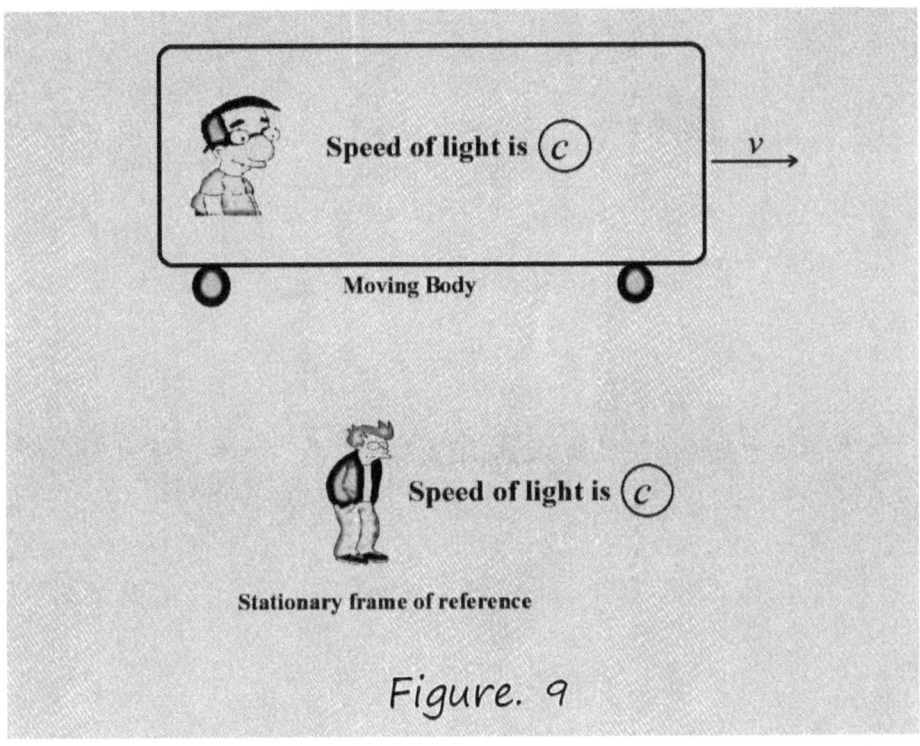

Figure. 9

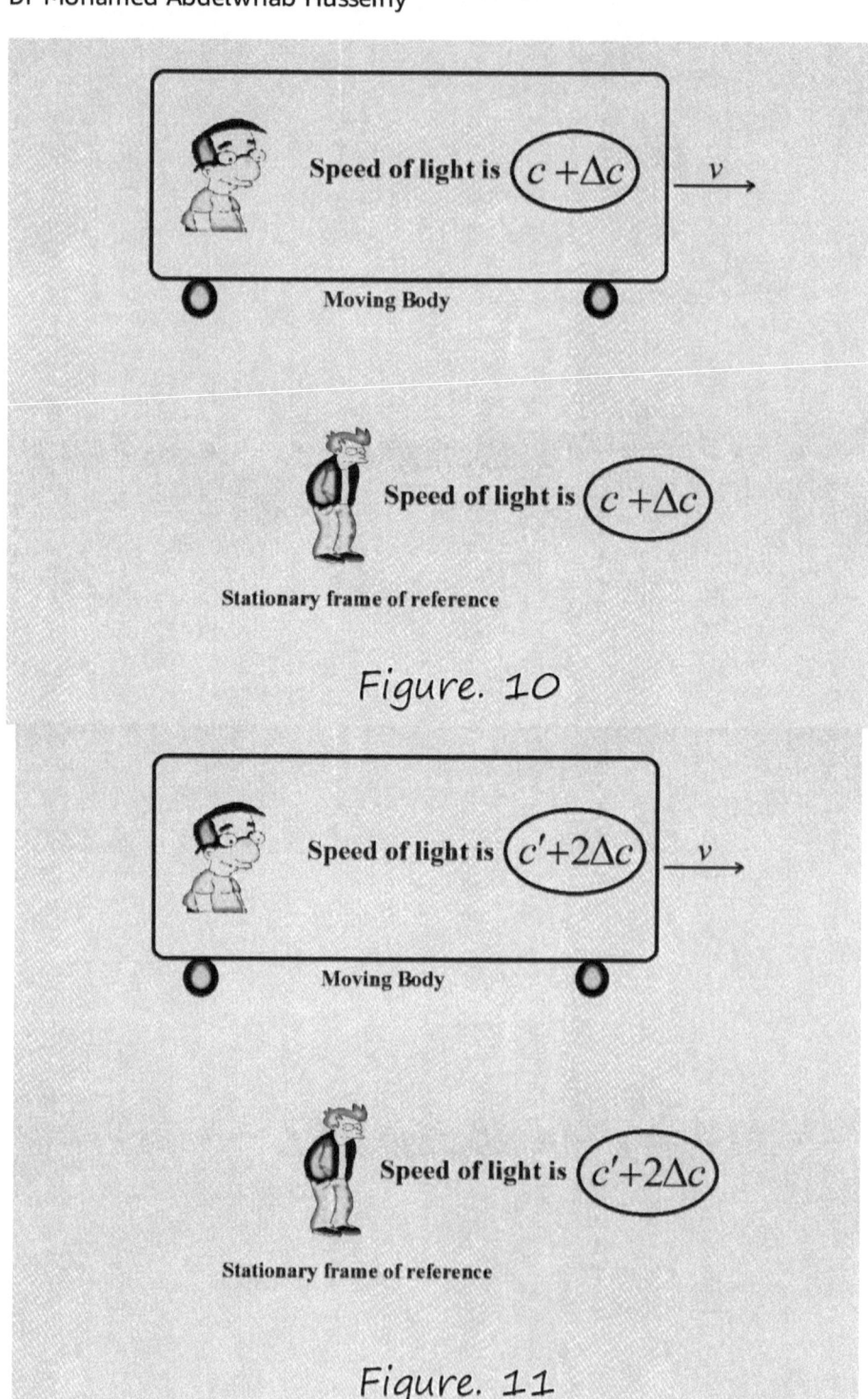

Figure. 10

Figure. 11

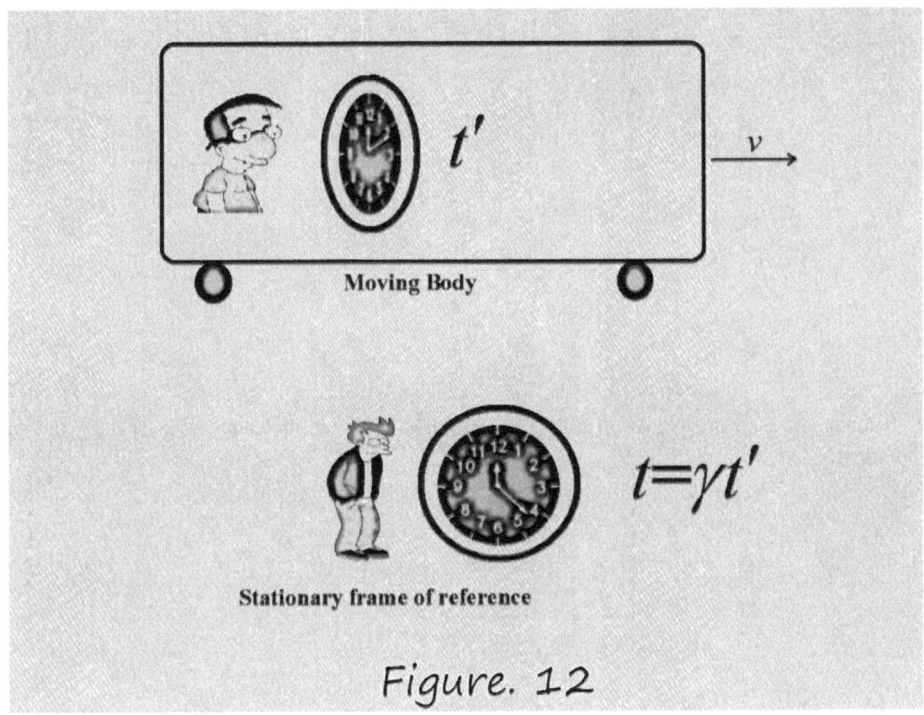

Figure. 12

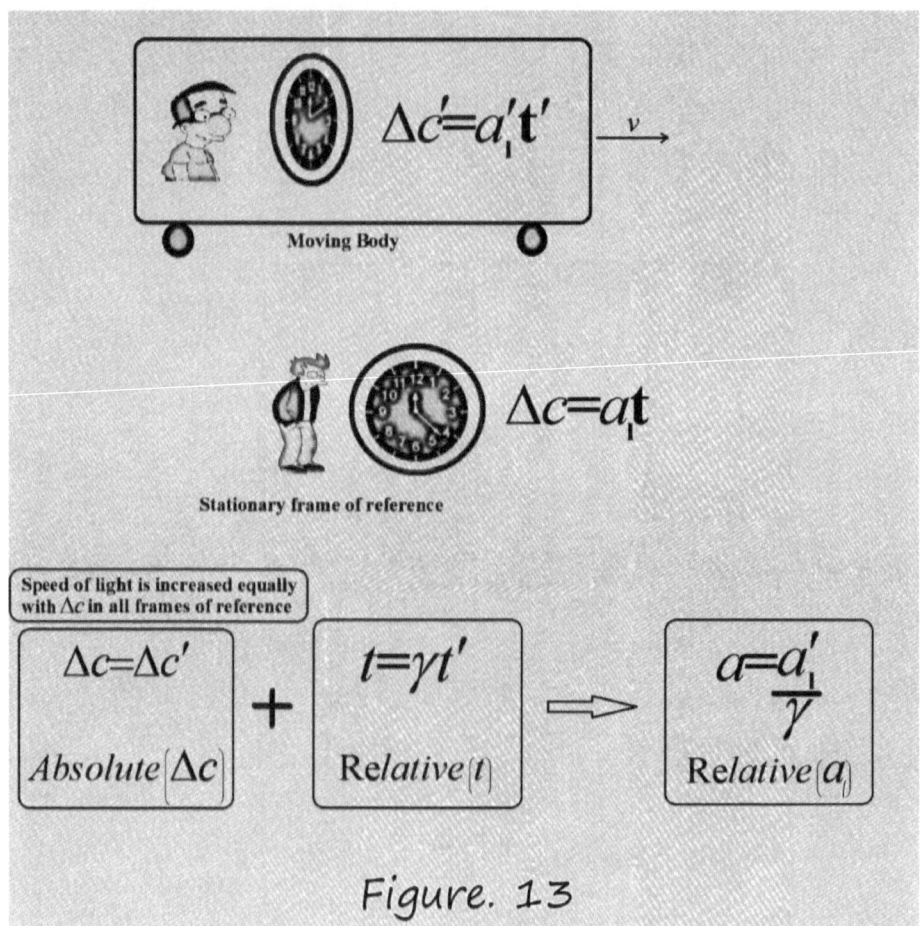

Figure. 13

* * *

CHAPTER 3:

Calculation the value of acceleration of light

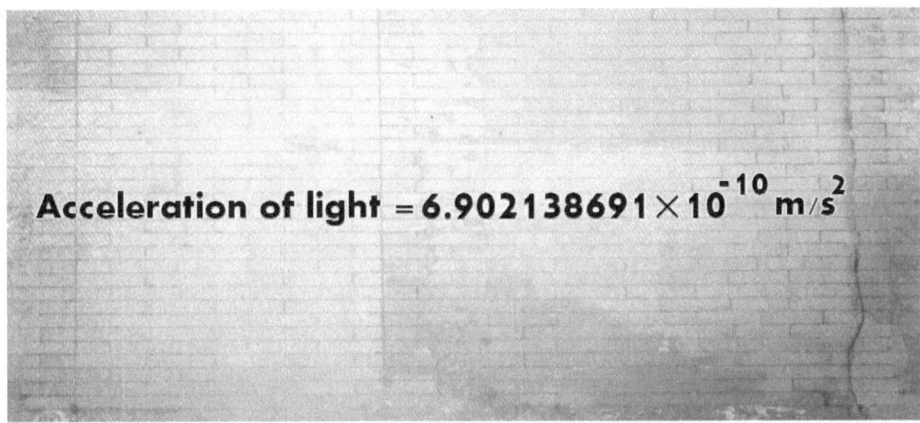

Acceleration of light $= 6.902138691 \times 10^{-10} \ m/s^2$

It is known that light requires time to travel a distance (x) and as the speed of light is increased with time, there is always a difference between speed of the emitted light (c_{emit}) and that of the observed light (c_{obsv}) where the speed of the emitted light is less than the speed of the observed light, as the following equations shown,

$$c_{obsv} \rangle c_{emit},$$
$$c_{obsv} = c_{emit} + \Delta c, (2)$$

where (Δc) denotes the change in speed of light along the space between the light source and the observer that receives the light (Fig. 14).

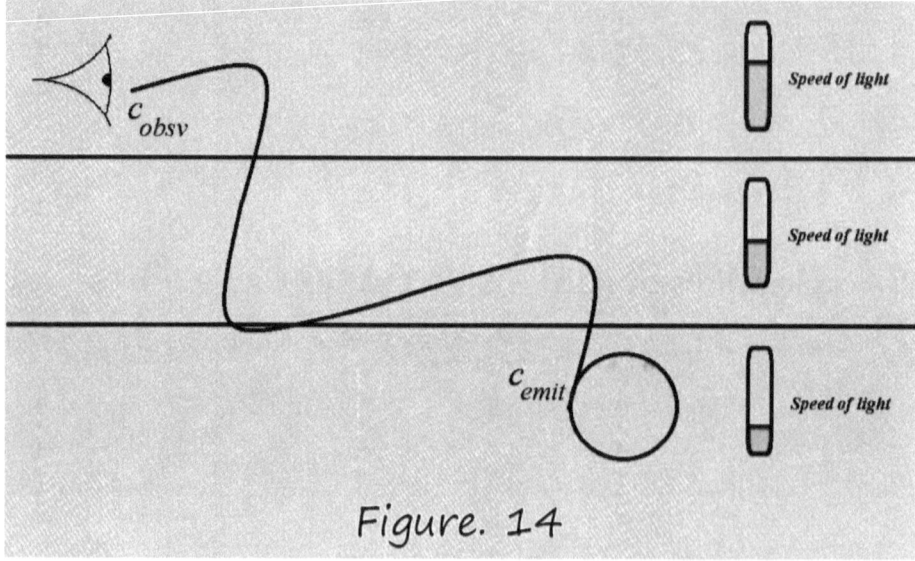

Figure. 14

If we consider that (Δc) denotes the change of speed of light, the acceleration of light (a_l) can be determined as follows:

$$acceleration \ of \ light = \frac{change \ in \ speed \ of \ light}{time}, (3)$$

$$a_l = \frac{\Delta c}{t} = \frac{c_{now} - c_{initial}}{t}, (4)$$

where ($c_{initial}$) denotes the speed of light at birth of the universe while (c_{now}) denotes the current speed of light in the universe (Fig. 15).

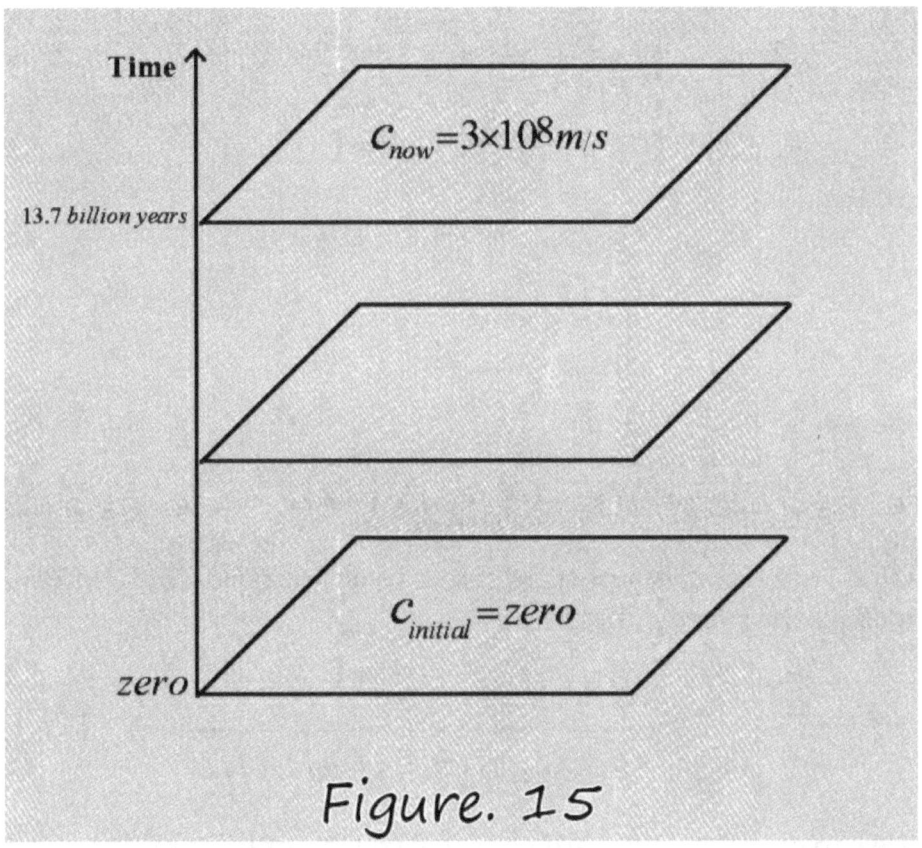

$c_{now} = 3 \times 10^8 m/s$

$c_{initial} = zero$

zero

13.7 billion years

Time

Figure. 15

If we consider that the time (t) is equal to the age of universe (A$_o$), the change in speed of light in this case is equal to the current speed of light as the speed of light at birth of the universe is equal to zero, as follows:

$$if \ t = A_o,$$

$$c_{initial} = zero$$

I find,

$$\Delta c = c_{now} - zero = c_{now},$$

Thus,

$$a_l = \frac{c}{A_o}, (5)$$

And as,

$$H_o = \frac{1}{A_o},$$

We get,

$$a_l = H_o.c, (6)$$

I find that Hubble's constant refers to acceleration of light divided by the speed of light:

$$H_o = \frac{a_l}{c} = \frac{acceleration \ of \ light}{speed \ of \ light}, (7)$$

We can calculate the value of acceleration of light by (Eq. 6) as follows:

$$a_l = H_o.c,$$

As Huble's constant equals,

$$H_o = 71 \ km/sec/mega \ parsec$$

I get,

$$a_l = \frac{71km}{s.mpc} \cdot \frac{3 \times 10^5 \ km}{s},$$

$$a_l = \frac{1km}{1mpc} \cdot \frac{71 \times 3 \times 10^5 \, km}{s^2},$$

$$a_l = \frac{1km}{3.086 \times 10^{19} \, km} \cdot \frac{71 \times 3 \times 10^5 \, km}{s^2},$$

$$a_l = \frac{1}{3.086 \times 10^{19}} \cdot \frac{213 \times 10^5 \, km}{s^2},$$

$$a_l = \frac{213 \times 10^5}{3.086 \times 10^{19}} \, km/s^2,$$

$$a_l = 6.902138691 \times 10^{-13} \, km/s^2$$

$$a_l = 6.902138691 \times 10^{-10} \, m/s^2, (8)$$

That value represents the acceleration of light in the temporary static universe, that value is constant where (a_l) doesn't change with the progress of time.

We must be noted that value of acceleration of light is relative where it is decreased (with motion) in moving frames or moving space with respect to the stationary frame of reference as we mentioned in the previous chapter, it is given by,

$$6.9 \times 10^{-10} \sqrt{1 - \frac{v^2}{c^2}},$$

where (v) is the velocity of the moving frame.

Using (Eq. 6), I get:

$$a_l = H_o.c,$$

$$a_l = H_o.\frac{D}{t},$$

$$a_l.t = H_o.D,$$

$$\Delta c = H_o.D, (9)$$

According to the previous equation, I find that (Δv) in Hubble's law (that is $\Delta v = H_o.D$) doesn't refers to the recession speeds of galaxies rather it refers to the difference in speed of light (Δc) between the observer and the observed body or in other words, it refers to the difference between speed of the emitted light and that of the observed light.

PART TWO:

Effects of acceleration of light on nature of T.S universe

I find that the acceleration of light affects the nature of the temporary static universe (T.S universe) in many properties that:

1. Wavelength of light
2. Time

3. *Momentum of light*
4. *The force*
5. *Mass*
6. *Gravitational constant*
7. *Einstein's gravitional constant*
8. *Cosmic microwave background radiation*
9. *Velocity, linear momentum and wavelength of bodies*

I show the affects of acceleration of light in details in the following chapters.

CHAPTER 1

Wavelength of light in T.S universe

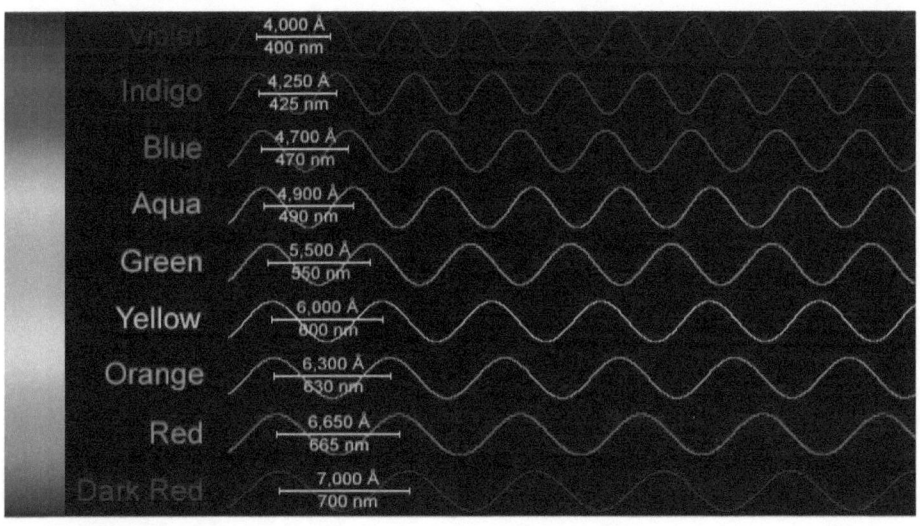

1.1 Wavelength of light is stretched with time "Redshift"

R edshift happens when the emitted light from the light source is increased in wavelength or shifted to the red end of the spectrum. As an object moves away from us, the light waves emitted by the source are stretched out, which makes them have a lower pitch and moves them towards the red end of the electromagnetic spectrum. In this case of light waves, this is called redshift. Redshift is considered as a proof of the universal expansion and thus a Big Bang (Fig. 16).

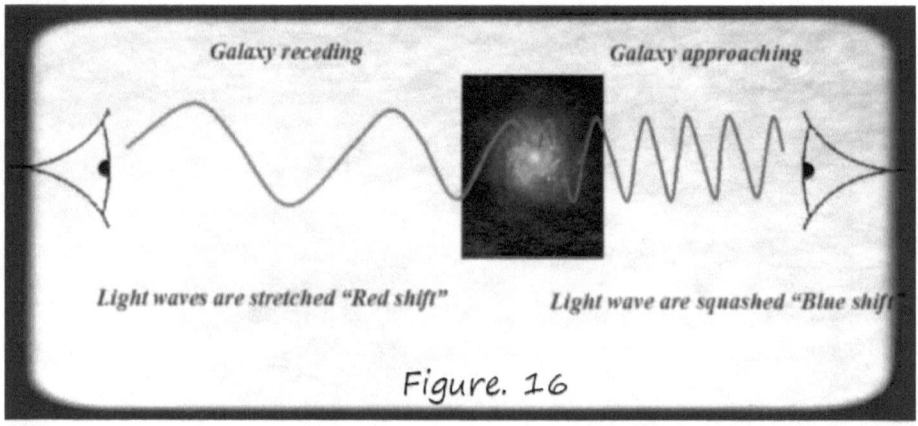

Figure. 16

Since the discovery of redshift-distance relationship by Hubble in 1929, there are alternative theories have been developed to explain the redshift of cosmological objects, such as tired light theory that is proposed that the redshift was caused by the photons losing energy as they passed through the matter or forces in intergalactic space. The theory has never been commonly accepted and seems to be now completely abandoned.

The redshift of a cosmological object can be computed by comparing the spectrum of the light arriving to us with its theoretical emission spectrum. The measured red shifts are usually stated in terms of (z) parameter that can be determined by the following equation:

$$z = \frac{\Delta\lambda}{\lambda_{emit}}, (10)$$

Where ($\Delta\lambda$) denotes the change of wavelength of light and (λ_{emit}) denotes the wavelength of the emitted light.

According to the mentioned postulate that shows the light is accelerated, we can explain why the wavelength of the emitted light is stretched in the static universe with respect to the observer.

In the static universe, the speed of light is increased with time and as light needs time to travels from the observed body to the observer, the speed of the observed light (c_{obsv}) will be more than the speed of the emitted light (c_{emit}) and as the energy of the photon doesn't change with time, the wavelength of light must be increased with time (Fig. 17) as the following equation shown:

$$E = hf = \frac{h.\uparrow c}{\uparrow \lambda}, (11)$$

$$E = hf = \frac{h.\Delta c}{\Delta\lambda}, (12)$$

Where (E) denotes the energy of photon that it doesn't change with time, (h) denotes Planck's constant and (f) denotes the frequency of light that it doesn't change with time.

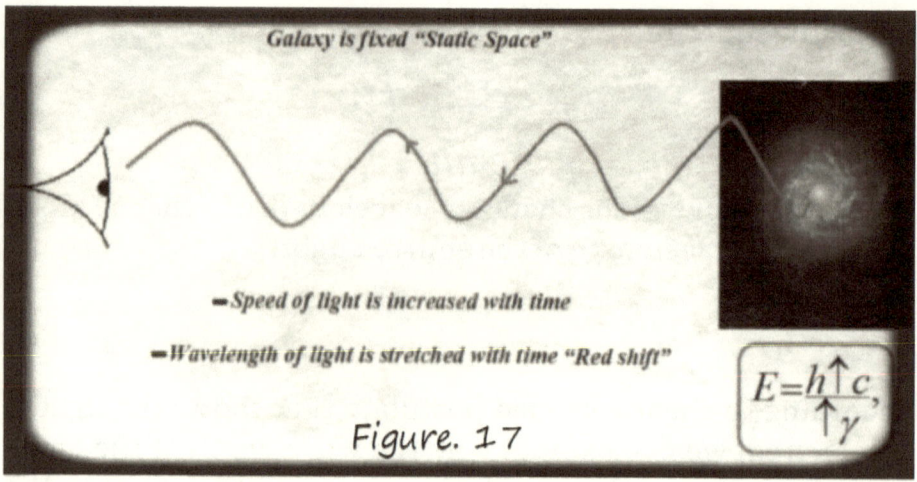

Figure. 17

From (Eq. 10), we can derive the relation between the wavelength of the emitted light and that of the observed light as,

$$z = \frac{\lambda_{obsv} - \lambda_{emit}}{\lambda_{emit}},$$

$$z = \frac{\lambda_{obsv}}{\lambda_{emit}} - 1,$$

$$1 + z = \frac{\lambda_{obsv}}{\lambda_{emit}},$$

$$\lambda_{obsv} = \lambda_{emit}(1+z), \quad (13)$$

Where (λ_{emit}) denotes the wavelength of the emitted light while (λ_{obsv}) denotes the wavelength of the observed light.

Therefore,

$$\frac{c_{obsv}}{f} = \frac{c_{emit}}{f}(1+z),$$

$$c_{obsv} = c_{emit}(1+z), \quad (14)$$

The previous equation shows the speed of the emitted light related to that of the observed light.

As the speed of light is increased with time, we can replace (c_{obsv}) that refers to the speed of the observed light with (c_{now}) that refers to the current speed of light in the static universe while (c_{emit}) that refers to the speed of emitted light can be replaced with (c_{past}) that refers to the speed of light in the static universe in the past (Fig. 18),

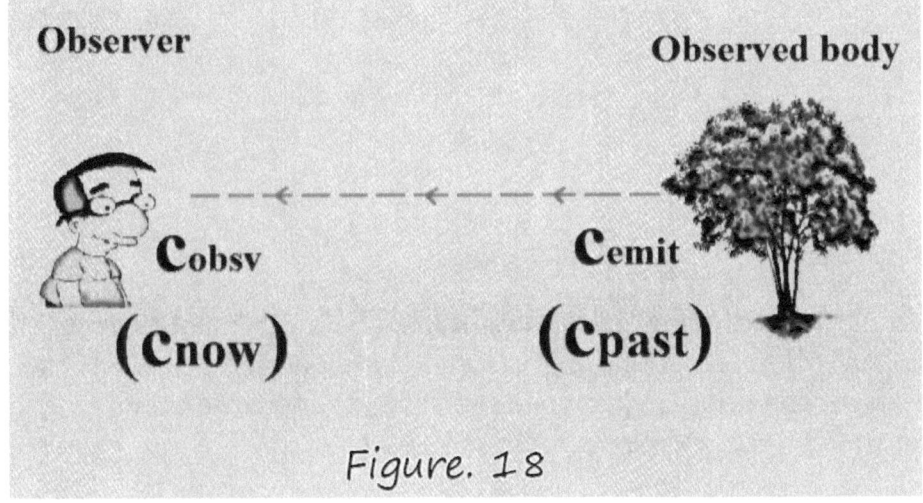

Observer   **Observed body**

Cobsv   **Cemit**

(Cnow)   **(Cpast)**

Figure. 18

Thus we can rewrite (Eq. 14) as,

$$c_{now} = c_{past}(1+z), (15)$$

Accordingly, acceleration of light and constancy of energy of light in the temporary static universe cause stretching of wavelength of light with time. If (λ_{now}) denotes the current wavelength of light while (λ_{past}) denotes the wavelength of light in the past, the relation between (λ_{now}) & (λ_{past}) can be given as follows:

$$\lambda_{now} = \lambda_{past}(1+z), (16)$$

Also we can determine the relation between the red shift (z)
and speed of light as follows:

$$\frac{c_{now}}{c_{past}} = 1+z,$$

$$\frac{c_{now}}{c_{past}} - 1 = z,$$

$$\frac{c_{now} - c_{past}}{c_{now}} = z,$$

$$z = \frac{\Delta c}{c_{past}}, (17)$$

Thus the more red shift (z) refers to the more change in speed
of light between the observer and the observed body. Also we
can get the relation between red shift (z) and time taken by the
emitted light (t_{past}) as follows:

$$z = \frac{\Delta c}{c_{past}} = \frac{H_o . D}{c_{past}} = H_o . t_{past}, (18)$$

* * *

1.2 The change of wavelength of light (Δλ)

A s the change of the speed of light (Δc) is absolute, I find that the change of wave length of light (Δλ) must be absolute also where the frequency of light (f) is still constant, as the following equation shows,

$$f = \frac{\Delta c}{\Delta \lambda},$$

"Wavelength of light is increased equally everywhere".

or in other words

"the change of wavelength of light (Δλ) is absolute".

The change of speed of light (Δc)	The change of wavelength of light (Δλ)
Absolute	Absolute

1.2.1 Mathematics of (Δλ)

In this section, I show the mathematical derivation of the change of stretching of wavelength of light (Δλ) in the temporary static universe as follows:

$$\lambda_{now} = \lambda_{past}\,(1+z),$$

$$\lambda_{now} = \lambda_{past} + \lambda_{past}\cdot z,$$

$$\lambda_{now} - \lambda_{past} = \lambda_{past}\cdot z,$$

$$\Delta\lambda = \lambda_{past}\cdot z,$$

From (Eq. 18), I find

$$\Delta\lambda = \lambda_{past}\left(H_\circ\cdot t_{past}\right),\,(19)$$

In the previous equation, the positive value of the change of wavelength of light ($\Delta\lambda$) shows that the wavelength of light is increased or stretched.

1.2.2 Another method to derive the change of wavelength of light ($\Delta\lambda$):

$$\Delta\lambda = \lambda_{now} - \lambda_{past},$$

From (Eq. 16) I find,

$$\Delta\lambda = \lambda_{now} - \frac{\lambda_{now}}{1+z},$$

$$\Delta\lambda = \lambda_{now}\left(1 - \frac{1}{1+z}\right),$$

And as,

$$z = \frac{\Delta c}{c_{past}},$$

I get,

$$\Delta \lambda = \lambda_{now} \left(1 - \frac{1}{1 + \dfrac{\Delta c}{c_{past}}} \right),$$

$$\Delta \lambda = \lambda_{now} \left(1 - \frac{1}{1 + \left(\dfrac{c_{now} - c_{past}}{c_{past}} \right)} \right),$$

$$\Delta \lambda = \lambda_{now} \left(1 - \frac{1}{1 + \left(\dfrac{c_{now}}{c_{past}} - 1 \right)} \right),$$

$$\Delta\lambda = \lambda_{now}\left(1 - \frac{1}{\left(\dfrac{c_{now}}{c_{past}}\right)}\right),$$

$$\Delta\lambda = \lambda_{now}\left(1 - \frac{c_{past}}{c_{now}}\right),$$

$$\Delta\lambda = \lambda_{now}\left(\frac{c_{now} - c_{past}}{c_{now}}\right),$$

$$\Delta\lambda = \lambda_{now}\left(\frac{\Delta c}{c_{now}}\right),$$

$$\Delta\lambda = \lambda_{now}\left(\frac{H_\circ.D}{\left(\dfrac{D}{t_{now}}\right)}\right),$$

$$\Delta\lambda = H_\circ\left(\lambda_{now}.t_{now}\right), (20)$$

From (Eq. 19&20), I find that:

$$\Delta\lambda = H_\circ\left(\lambda_{now}.t_{now}\right) = H_\circ\left(\lambda_{past}.t_{past}\right), (21)$$

The unit of $(\Delta\lambda)$ can be obtained as follows:

$$\Delta\lambda = s^{-1}(m.s) = m,$$

Thus the unit of $(\Delta\lambda)$ is the meter.

* * *

1.3 Speed of elongation of wavelength of light (v_λ)

In physics, acceleration is the change of velocity of an object with time. In part one, I discussed why speed of light is changed or increased and I calculate the value of the change of speed of light with time (a_l) that is called acceleration also. In this sction, I am talking about something like that.

1.3.1 Mathematics of (v_λ)

We know that wavelength of light is stretched ($\Delta\lambda$) that is given by (Eq. 19&20). In this section, I show the mathematical derivation of the change of wavelength of light with time (v_λ), I called this property as "speed of elongation of wavelength of light".

Speed of elongation of wavelength of light (v_λ):
it refers to the rate of stretching "elongation" of wavelength of light, the rate of change of wavelength of light, or the change of wavelength

of light with time.

Thus, (v_λ) can be given as follows:

$$v_\lambda = \frac{\Delta\lambda}{t_{now}}, (22)$$

From (Eq. 20), I find

$$v_\lambda = \frac{H_\circ.\lambda_{now}.t_{now}}{t_{now}},$$

$$v_\lambda = H_\circ.\lambda_{now}, (23)$$

where the unit of (v_λ) can be obtained as follows:

$$v_\lambda = s^{-1}.m = m/s,$$

1.3.2 Elongation of wavelength of light has a constant relativistic speed:

In part one, I show that the acceleration of light is constant and its value equals (6.9×10-10m/s2), that value remains constant with the progress of time, rather it can be changed as a result of time relativity between inertial frames where "the acceleration of light is relativistic between inertial frames".

As the light is accelerating with a constant value, I find that the speed of elongation of wavelength of light per the second must be constant also,

"The speed of elongation of wavelength of light per the second is the same for each second".

I find that speed can be changed only with motion where its value in moving spaces is decreased from perspective of observers at rest due to relativity of time where "the speed of elongation of wavelength of light is relativistic between inertial frames".

Acceleration of light (a_λ)	Speed of elongation of wavelength of light (v_λ)
Relative	Relative

Accordingly, if (v_λ') denotes speed of elongation of wavelength of light in the moving frame with respect to the moving observer and (v_λ) denotes speed of elongation of wavelength of light in the moving frame with respect to the stationary frame of reference, the relation between (v_λ') & (v_λ) is given as:

$$v_\lambda = \frac{\Delta\lambda}{t_{now}},$$

And as time is dilated with motion with respect to the observer at rest where,

$$t_{now} = \gamma t'_{now},$$

I get,

$$v_\lambda = \frac{\Delta\lambda}{\gamma \cdot t'_{now}},$$

As ($\Delta\lambda$) is similar to (Δc) where the two quantities are absolute or the same for all frames of refernce, I find that (v_λ) must be given as,

$$\frac{v_\lambda{}'}{\gamma} = \frac{\Delta\lambda}{\gamma \cdot t'_{now}},$$

$$v_\lambda = \frac{v_\lambda{}'}{\gamma}, (24)$$

Regarding to (Eq. 24), the speed of elongation of wavelength of light in a moving space is decreased with respect to the observer at rest.

1.3.3 Relation between speed of elongation of wavelength of light (v_λ) and acceleration of light (a_l)

We can find the relation between (v_λ) and (a_l) as follows:

$$v_\lambda = H_o \cdot \lambda_{now},$$

$$v_\lambda = \frac{a_l}{c_{now}} \cdot \lambda_{now},$$

$$v_\lambda = \frac{a_l}{f}, (25)$$

We can obtain a single universal equation includes the absolute change of speed of light (Δc), the absolute change of wavelength of light ($\Delta\lambda$), the relative acceleration of light (a_l) and the relative speed of elongation of wavelength of light (v_λ) as,

$$v_\lambda = \frac{a_l.\Delta\lambda}{\Delta c}, (26)$$

* * *

CHAPTER 2

Time in T.S universe

2.1 Contraction of time

T he decreasing in time of the temporary static universe refers to contraction of time where time moves faster and light travels faster in that static universe. The contraction of time means that the second in any frame moves faster with respect to the previous one in the same frame.

Accordingly If (t_{now}) denotes the current second in a light's clock while (t_{past}) denotes the second in the same light's clock in the past, the relation between (t_{now}) & (t_{past}) can be given as follows:

$$c_{now} = c_{past}(1+z),$$

$$\frac{D}{t_{now}} = \frac{D}{t_{past}}(1+z),$$

$$t_{now} = \frac{t_{past}}{1+z}, \quad (27)$$

The previous equation shows that time in the static universe was dilated in the past with respect to the current time.

As I show in the previous chapter, the relation between red shift (z) and time taken by the emitted light (t_{past}) is given as,

$$z = \frac{\Delta c}{c_{past}} = \frac{H_o.D}{c_{past}} = H_o.t_{past},$$

while the relation between redshift (z) and current time (t_{now})

can be given as:

$$1 + z = 1 + \frac{\Delta c}{c_{past}},$$

$$1 + z = 1 + \frac{\Delta c}{c_{now} - \Delta c},$$

$$1 + z = \frac{(c_{now} - \Delta c) + \Delta c}{c_{now} - \Delta c},$$

$$1 + z = \frac{c_{now}}{c_{now} - \Delta c},$$

$$1 + z = \frac{1}{\dfrac{c_{now} - \Delta c}{c_{now}}},$$

$$1 + z = \frac{1}{1 - \dfrac{\Delta c}{c_{now}}}, \quad (28)$$

$$1+z = \cfrac{1}{1 - \cfrac{a_l \cdot t_{now}}{c_{now}}},$$

And as,

$$H_o = \frac{a_l}{c_{now}},$$

I find that,

$$1+z = \frac{1}{1-\left(H_o \cdot t_{now}\right)}, (29)$$

Regarding to the following three equations, that are:

-(Eq. 15) that shows the relation between (c_{now}) and (c_{past}).

-(Eq. 18) that shows the relation between redshift (z) and time taken by the emitted light (t_{past}).

-(Eq. 29) that shows the relation between redshift (z) and time taken by the observed light (t_{now}).

I get the following two equations,

$$c_{now} = c_{past}\left(1+z\right) = c_{past}\left(1+H_o \cdot t_{past}\right), (30)$$

$$c_{past} = \frac{c_{now}}{1+z} = c_{now}\left(1-H_o \cdot t_{now}\right), (31)$$

* Problems

Problem one:

If time interval between two bodies (A) and (B) is equal to (0.1 sec), what is the time interval between them when speed of light was equal to $(3 \times 10^2 m/s)$?
use the current speed of light $(c_{now}) = 3 \times 10^8 m/s$

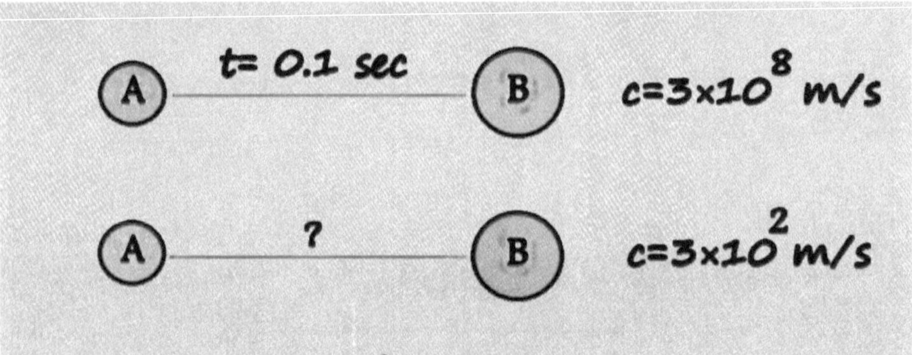

t= 0.1 sec $c = 3 \times 10^8$ m/s

? $c = 3 \times 10^2$ m/s

Figure. 19

Solution:

$$t_{past} = t_{now}(1+z),$$

And as,

$$1+z = \frac{c_{now}}{c_{past}},$$

We get,

$$t_{past} = t_{now}\left(\frac{c_{now}}{c_{past}}\right),$$

$$t_{past} = 0.1 \left(\frac{3 \times 10^8}{3 \times 10^2} \right),$$

$$t_{past} = 0.1 \times 1 \times 10^6 = 1 \times 10^5 \text{ sec,}$$

So time interval between the two bodies (a & b) was equal to 1×10^5 sec when speed of light is equal to 3×10^2 m/s.

Problem two:

If the current speed of light in a light's clock is equal to (3×10^8 m/s), what is the speed of light in that clock after (5 sec)?

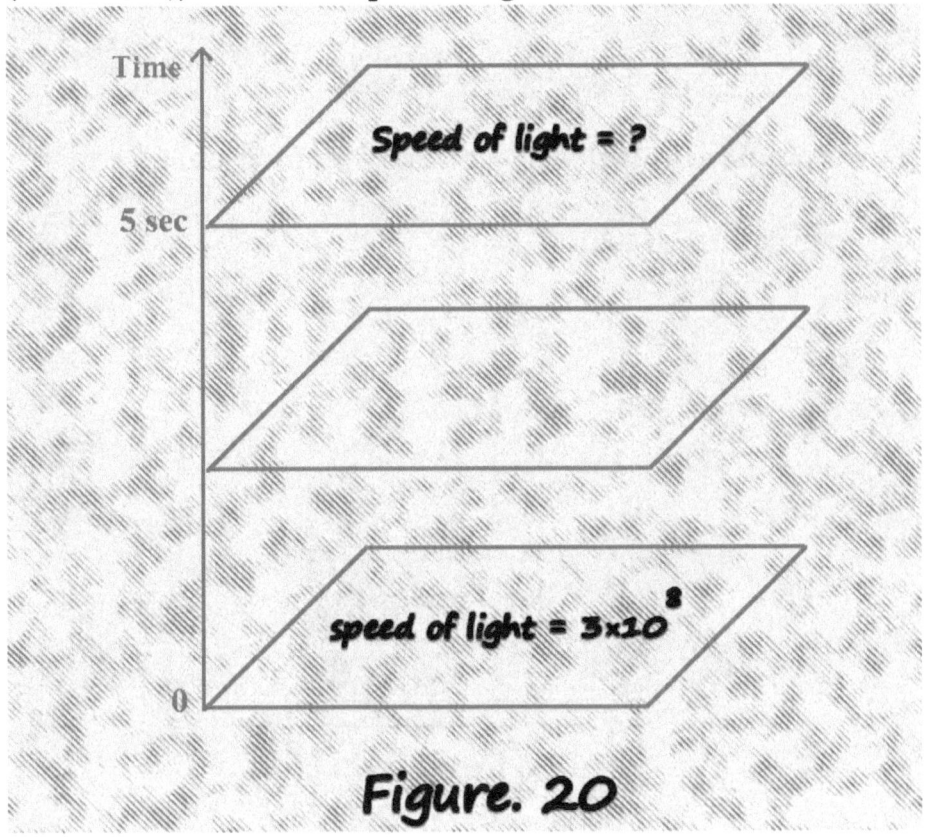

Figure. 20

Solution:

-As the speed of light is increased with time, the current value of speed of light that is equal to (3×108m/s) or (c_{now}) will be increased to be equal (c_{future}), and as the relation between (c_{now}) & (c_{past}) is given as,

$$c_{now} = c_{past}(1+z),$$

the relation between (c_{now}) & (c_{future}) can be given as,

$$c_{future} = c_{now}(1+z),$$

Due to contraction of time in the static universe, the presence of word "after" in "after 5 sec" in this problem means that time (5 sec) is related to the future's perspective or the observer perspective, so:

$$t_{future} = 5\sec$$

-By the previous knowledge, we can solve this problem as,

$$c_{future} = c_{now}(1+z),$$

And as,

$$1+z = \frac{1}{1 - H_o.t_{future}},$$

I get,

$$c_{future} = \frac{c_{now}}{1 - H_o.t_{future}},$$

$$c_{future} = \frac{3 \times 10^8}{1 - \left(\left(2.3 \times 10^{-18}\right) \times 5\right)},$$

$$c_{future} = 300000000.00000000345\,m/s,$$

The previous value is the speed of light after 5 sec.

* * *

2.2 The change of time (Δt)

A s the change of speed of light (Δc) is absolute and space is static, the change of time (Δt) must be absolute also where the distance that the light travels (D) is constant, I find

"Time is decreased or contracted equally everywhere".

or in other words

"The change of time (Δt) is absolute".

The change of speed of light (Δc)	The change of time (Δt)
Absolute	Absolute

2.2.1 Mathematical derivation of the change of time (Δt)

In this section, I show the mathematical derivation of the change of time (Δt) in the temporary static universe as follows:

$$t_{past} = t_{now}(1+z),$$

$$t_{past} = \frac{t_{now}}{1 - H_o.t_{now}},$$

$$t_{past} = \frac{1}{\dfrac{1 - H_o.t_{now}}{t_{now}}},$$

$$t_{past} = \frac{1}{\dfrac{1}{t_{now}} - H_o},$$

$$\frac{1}{t_{now}} - H_o = \frac{1}{t_{past}},$$

$$\frac{1}{t_{now}} - \frac{1}{t_{past}} = H_o,$$

$$\frac{t_{past} - t_{now}}{t_{now}.t_{past}} = H_o,$$

$$\frac{\Delta t}{t_{now}.t_{past}} = H_o,$$

$$\Delta t = H_\circ \left(t_{now} . t_{past} \right), (32)$$

2.2.2 Another method to get the change of time (Δt):

Also we can derive the same previous equation by another method as follows:

$$c_{now} . t_{now} = c_{past} . t_{past} = D,$$

$$c_{now} . t_{now} = D,$$

$$\left(c_{past} + \Delta c \right) \left(t_{past} - \Delta t \right) = D,$$

$$t_{past} - \Delta t = \frac{D}{c_{past} + \Delta c},$$

$$\Delta t = t_{past} - \frac{D}{c_{past} + \Delta c},$$

$$\Delta t = t_{past} - \frac{c_{past} . t_{past}}{c_{past} + \Delta c},$$

$$\Delta t = t_{past}\left(1 - \frac{c_{past}}{c_{past} + \Delta c}\right),$$

$$\Delta t = t_{past}\left(1 - \frac{1}{1 + \dfrac{\Delta c}{c_{past}}}\right),$$

$$\Delta t = t_{past}\left(1 - \frac{1}{1 + \dfrac{c_{now} - c_{past}}{c_{past}}}\right),$$

$$\Delta t = t_{past}\left(1 - \frac{1}{1 + \left(\dfrac{c_{now}}{c_{past}} - 1\right)}\right),$$

$$\Delta t = t_{past}\left(1 - \frac{1}{1+(1+z-1)}\right),$$

$$\Delta t = t_{past}\left(1 - \frac{1}{1+z}\right),$$

$$\Delta t = t_{past}\left(\frac{(1+z)-(1)}{1+z}\right),$$

$$\Delta t = t_{past}\left(\frac{z}{1+z}\right),$$

$$\Delta t = t_{now}.z,$$

$$\Delta t = t_{now}\left(H_{\circ}.t_{past}\right),$$

2.2.3 Does the change of time (Δt) is increased or decreased?

$$\Delta t = t_{now} - t_{past},$$

$$\Delta t = t_{now} - \left(t_{now}(1+z)\right),$$

$$\Delta t = t_{now} - t_{now} - t_{now}.z,$$

$$\Delta t = -t_{now} \cdot z,$$

$$\Delta t = -H_{\circ} \left(t_{now} \cdot t_{past} \right),$$

The negative value of the change of time (Δt) shows that time is decreased or contracted. The unit of (Δt) can be obtained as follows:

$$\Delta t = s^{-1} .s.s = s,$$

Thus the unit of (Δt) is the second.

❊ ❊ ❊

2.3 Rate of contraction of time "cosmological red-shift"

2.3.1 Mathematics of (z)

We know that time is changed or contracted (Δt) as it is shown in (Eq. 32). In this section, I show the mathematical derivation of the change of time with time or the rate of contraction of time.

It can be given as,

$$x = \frac{\Delta t}{t_{now}}, (33)$$

where (x) denotes the rate of contraction of time, using (Eq. 32), I get

$$x = \frac{H_{o}\left(t_{now} \cdot t_{past}\right)}{t_{now}},$$

$$x = H_{o} \cdot t_{past}, (34)$$

From (Eq.18) I find that,

$$x = z, (35)$$

I find that the rate of contraction of time or the change of time with time (x) is refer to the cosmological redshift (z).

2.3.2 The contraction of time per the second has a constant re-lativistic value:

As the light is accelerate with a constant value, I find that the contraction of time per one sec must be constant also and by this way, space is still static, I mean that;

"The amount of contraction of time per the second is the same for each second".

I find that amount can be changed only with motion where its value in moving spaces is decreased from persepctive of observers at rest due to relativity of time where "the rate of contraction of time is relativistic between inertial frames".

Acceleration of light (a_L)	The rate of contraction of time (z)
Relative	Relative

If (z′) denotes the rate of contraction of time of the moving frame with respect to the moving observer while (z) denotes the rate of contraction of time of the moving frame with respect to the stationary frame of reference, the relation between (z′) & (z) is given as:

$$z = \frac{\Delta t}{t_{now}},$$

And as time is dilated with motion with respect to the observer at where,

$$t_{now} = \gamma t'_{now},$$

I get,

$$z = \frac{\Delta t}{\gamma \cdot t'_{now}},$$

As the change of time (Δt) is absolute or the time is contracted equally in all frames of reference, I find that (z) must be given as,

$$z = \frac{z'}{\gamma}, (36)$$

Regarding to (Eq. 36), the rate of contraction of time is relative regarding to motion where it is decreased with respect to the stationary frame of reference.

2.3.3 Relation between the rate of contraction of time (z) and acceleration of light (a_l)

We can find the relation between (z) and (a_l) using (Eq. 17) as follows:

$$z = \frac{\Delta c}{c_{past}},$$

Thus,

$$z = \frac{a_1 . t_{now}}{c_{past}}, (37)$$

We can obtain the relation between (z) and (a_1) in another form as follows:

$$z = H_\circ . t_{past}$$

$$z = \frac{a_1}{c_{now}} t_{past}, (38)$$

the two equations (37 & 38) are valid because,

$$\frac{t_{now}}{c_{past}} = \frac{t_{past}}{c_{now}},$$

$$c_{now} . t_{now} = c_{past} . t_{past} = D,$$

where (D) is the distance that light travels, that is doesn't change with time.

2.4 Calculation of the value of contraction of time per one sec (z_1)

In this section, I will calculate the contraction of time per one sec, or in other words; the constant value of the change of time during the second.

Using (Eq. 29), I find,

$$z = \frac{1}{1-\left(H_o.t_{now}\right)} - 1,$$

$$z = \frac{1-\left(1-\left(H_o.t_{now}\right)\right)}{1-\left(H_o.t_{now}\right)},$$

$$z = \frac{H_o.t_{now}}{1-\left(H_o.t_{now}\right)},$$

$$z = \cfrac{1}{\left(\cfrac{1}{H_o.t_{now}}\right) - 1} , (39)$$

If (z_1) denotes the contraction of time per one sec (where t_{now} is equal to one sec), the value of (z_1) can be calculated as,

$$z_1 = \cfrac{1}{\cfrac{1}{H_o \times 1\,sec} - 1} = 2.3007 \times 10^{-18},$$

thus time is contracted with $2.3 \times 10{-}18$ per one sec,

As the contraction of time per one sec is equal to $2.3 \times 10{-}18$ and as the amount of contraction of time per the second is the same for each second, I find that the contraction of time per two sec must be equal to,

$$z = 2z_1 = 2\left(2.3007 \times 10^{-18}\right) = 4.6 \times 10^{-18},$$

To ensure the validity of (Eq. 39), we make (z) is equal to (2 sec) to discover if the equation will give $(4.6 \times 10{-}18)$ or not, I find that the (Eq. 39) gives the same expected result, as follows:

$$z = \cfrac{1}{\cfrac{1}{H_o \times 2\,sec} - 1} = \cfrac{1}{\cfrac{1}{2\left(2.3007 \times 10^{-18}\right)} - 1},$$

$$z = 4.6 \times 10^{-18},$$

* * *

2.5 Dilation of time required for observation "Dilation of time intervals"

Acceleration of light makes difference between speed of the emitted light (c_{emit}) and that of the observed light (c_{obsv}). Thus the average speed of light can be determined as follows:

$$c_{average} = \frac{c_{obsv} + c_{emit}}{2},$$

$$c_{average} = \frac{c_{obsv} + (c_{obsv} - \Delta c)}{2},$$

$$c_{average} = \frac{c_{obsv} + c_{obsv} - \Delta c}{2},$$

$$c_{average} = \frac{2c_{obsv} - \Delta c}{2},$$

$$c_{average} = \frac{2\left(c_{obsv} - \dfrac{\Delta c}{2}\right)}{2},$$

$$c_{average} = c_{obsv} - \frac{\Delta c}{2},$$

$$c_{average} = c_{obsv} - 0.5\Delta c, \quad (40)$$

Using (Eq. 40), we can determine the average speed of light.

Thus the light doesn't travel with the observed speed or the current speed that is $(3 \times 108 m/s)$ rather it travels with the average speed that is less than the observed speed of light where the required time for light to travel from the observed body to the observer (Fig. 21) is given as,

$$t_{interval} = \frac{D}{c_{average}},$$

$$t_{interval} = \frac{D}{c_{obsv} - 0.5\Delta c}, \quad (41)$$

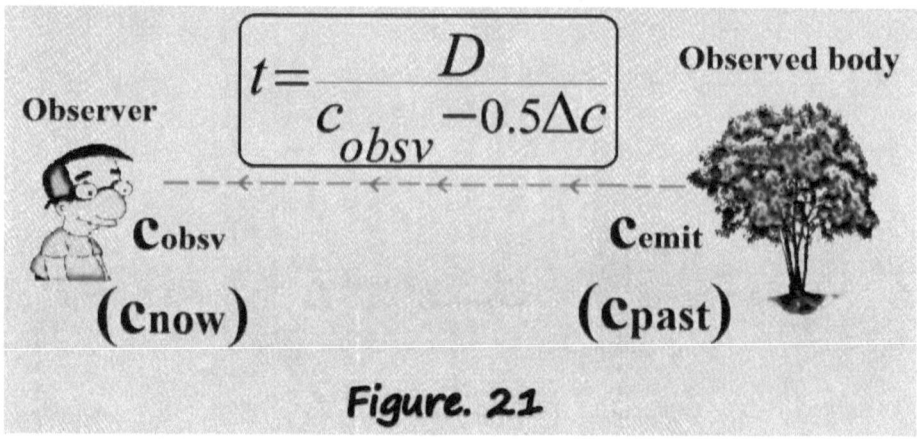

Figure. 21

Therefore, the relation between the time interval and the time of the observer is given as,

$$c_{average} . t_{interval} = D = c_{obsv} . t_{obsv} = c_{emit} . t_{emit},$$

$$c_{average} . t_{interval} = c_{obsv} . t_{obsv},$$

$$\left(c_{obsv} + 0.5\Delta c \right) t_{interval} = c_{obsv} . t_{obsv},$$

$$\left(1 + \frac{0.5\Delta c}{c_{obsv}} \right) t_{interval} = t_{obsv},$$

$$\left(1 + \frac{0.5 H_\circ . D}{\left(\dfrac{D}{t_{obsv}} \right)} \right) t_{interval} = t_{obsv},$$

$$\left(1+0.5H_\circ.t_{obsv}\right)t_{interval} = t_{obsv},$$

$$t_{interval} = \frac{t_{obsv}}{1+0.5H_\circ.t_{obsv}},$$

$$t_{interval} = \frac{1}{\dfrac{1}{t_{obsv}}+0.5H_\circ},(42)$$

* * *

2.6 Life span of time intervals "Expired time intervals"

I find that the phenomena of contraction of time in T.S universe causes necessary death or disappearnce of time intervals between bodies after certain duration (Fig. 22).

Time death: it is the necessary end that at it, time interval between two bodies is contracted to reach to the level of disappearance or the death (equal to zero).

Life span of time interval (t_{life}): it is the required duration for contraction of time interval between two bodies to be equal to zero "dead".

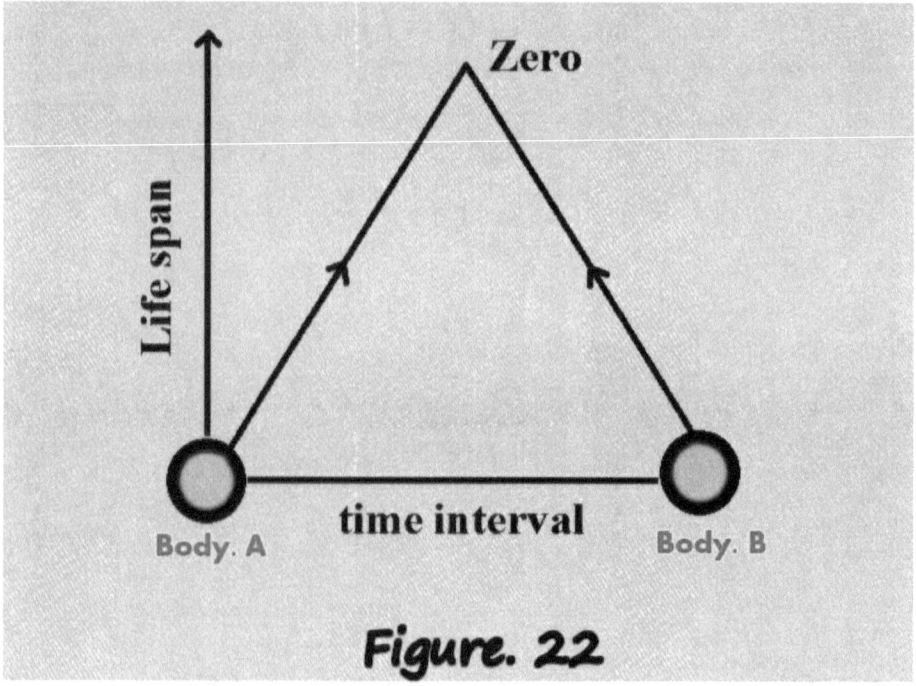

Figure. 22

Mathematics of (t_{life}):

If ($t_{interval}$)now denotes the current time interval between two bodies, and (z_1) denotes the contraction of time per the second, the life span of time interval (t_{life}) can be given as follows:

$$t_{life} = \frac{The\ current\ time\ interval}{The\ contraction\ of\ time\ per\ the\ seond},$$

$$t_{life} = \frac{\left(t_{interval}\right)_{now}}{z_1}, (43)$$

where,

$$z_1 = 2.3 \times 10^{-18}$$

The unit of (t_{life}) is (sec) as it is shown,

$$t_{life} = \frac{sec}{z_1} = sec,$$

It is easily to find that,

$$\left(t_{interval}\right)_{now} = \frac{\left(t_{interval}\right)_{past}}{1+z},$$

Problems:

Problem one:

If the current time interval between two bodies (A & B) is (5 sec), calculate the life span for this time interval. (or calculate the required duration for contraction of this time interval to be equal to zero, or When this time interval dies or disapears or completely contracted)?

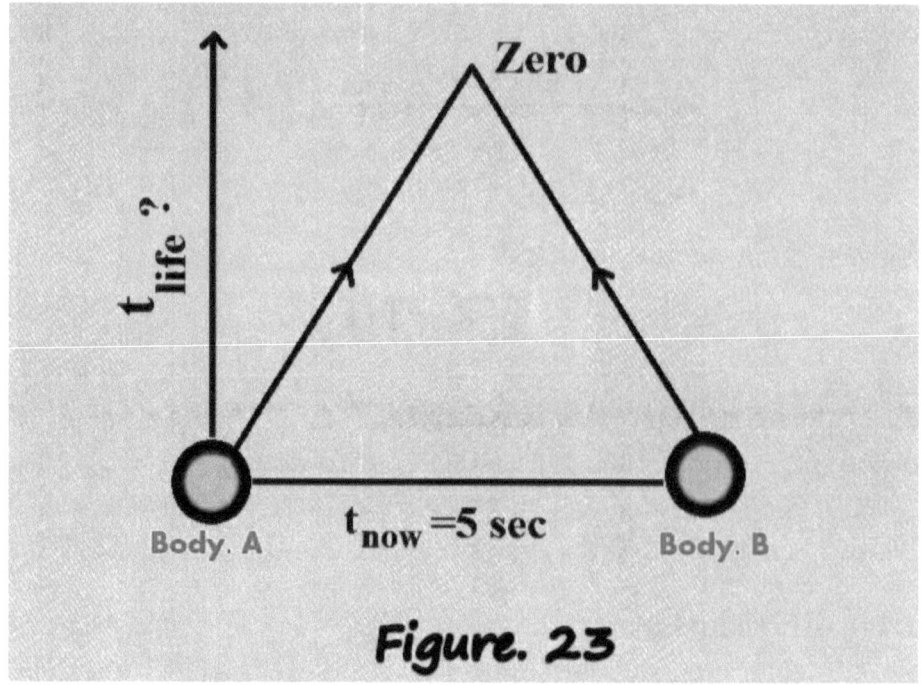

Figure. 23

<u>Solution:</u>

$$t_{life} = \frac{(t_{interval})_{now}}{z_1},$$

$$t_{life} = \frac{5}{2.3007 \times 10^{-18}},$$

$$t_{life} = 2173913043478260869 \text{ sec,}$$

where (z_1) that equals (2.3007×10^{-18}) is the value of contraction of time during the second.

<u>Problem two:</u>

If the distance between two bodies (A & B) is (8 meters), calculate the life span for time interval between the two bodies?

-use the current speed of light $(c_{now}) = 3 \times 108 m/s$

Solution:

$$D = c_{now} \cdot t_{now},$$

$$t_{now} = \frac{D}{c_{now}} = \frac{8}{3 \times 10^8} = 2.6 \times 10^{-8} \text{ sec},$$

And as,

$$\left(t_{interval}\right)_{now} = \frac{1}{\dfrac{1}{t_{now}} + 0.5H_o},$$

I get,

$$\left(t_{interval}\right)_{now} = \frac{1}{\dfrac{1}{2.6 \times 10^{-8}} + 0.5H_o},$$

$$\left(t_{interval}\right)_{now} = 2.6 \times 10^{-8} \text{ sec},$$

So,

$$t_{life} = \frac{\left(t_{interval}\right)_{now}}{z_1} = \frac{2.6 \times 10^{-8}}{2.3007 \times 10^{-18}},$$

$$t_{life} = 1.13 \times 10^{10} \text{ sec},$$

Thus the life span for time interval between the two bodies is equal to (1.13×1010) sec.

Briefly,

> *In T.S. universe; there is a life span for any time interval existing between two bodies, after the end of that life span, these two bodies will be connected with each other without require time forever.*

*   *   *

CHAPTER 3

Momentum of light in T.S universe

3.1 Momentum of light is decreased with time

T he acceleration of light and conservation of energy of light in the temporary static universe lead to decreasing in momentum of light with time. I find that the momentum of light in T.S universe is decreased with time, thus if (p_{now}) denotes the current momentum of light or momentum of the observed light while (p_{past}) denotes the momentum of light in the past or momentum of the emitted light, the relation between (p_{now}) & (p_{past}) can be given as follows:

$$E_{now} = E_{past} = E,$$

$$p_{now} \cdot c_{now} = p_{past} \cdot c_{past},$$

$$p_{now} \cdot c_{past} (1+z) = p_{past} \cdot c_{past},$$

$$p_{now} (1+z) = p_{past},$$

$$p_{now} = \frac{p_{past}}{1+z}, (44)$$

Another method to determine the relation between the current momentum of light and its momentum in the past:

From de Broglie's equation that is,

$$p = \frac{h}{\lambda},$$

-(h) is Planck's constant.

we can get the relation between the current momentum of photon and its momentum in the past (Eq. 44) by another way as follows:

$$h_{now} = h_{past} = h,$$

$$p_{now} \cdot \lambda_{now} = p_{past} \cdot \lambda_{past},$$

$$p_{now} \cdot \lambda_{past} (1+z) = p_{past} \cdot \lambda_{past},$$

$$p_{now} = \frac{p_{past}}{1+z},$$

* * *

3.2 The change of momentum of light (Δp)

As the change of speed of light is absolute and the energy is conserved, momentum of light must be absolute also. I find,

"Momentum of light is decreased equally everywhere".

or in other words

"The change of momentum of light (Δp) is absolute".

The change of speed of light (Δc)	The change of momentum (Δp)
Absolute	Absolute

3.2.1 Mathematical derivation of the change of momementum of light (Δp)

In this section, I show the mathematical derivation of the change of momentum (Δp) as follows:

$$\Delta p = p_{now} - p_{past}, (45)$$

$$\Delta p = p_{now} - \left(p_{now}\left(1+z\right)\right),$$

$$\Delta p = p_{now} - \left(p_{now} + \left(p_{now}\cdot z\right)\right),$$

$$\Delta p = p_{now} - p_{now} - \left(p_{now}\cdot z\right),$$

$$\Delta p = -p_{now} \cdot z,$$

as,

$$z = \frac{\Delta c}{c_{past}} = \frac{H_{\circ} \cdot D}{c_{past}} = H_{\circ} \cdot t_{past},$$

I get,

$$\Delta p = -p_{now} \left(H_{\circ} \cdot t_{past} \right), (46)$$

$$\Delta p = -p_{now} \left(H_{\circ} \cdot t_{now} \left(1 + z \right) \right),$$

$$\Delta p = -p_{past} \left(H_{\circ} \cdot t_{now} \right), (47)$$

The change of momentum of light can be given in another form using (Eq. 39) as,

$$\Delta p = -p_{now} \cdot z,$$

$$\Delta p = -p_{now} \left(\frac{1}{\frac{1}{H_{\circ} t_{now}} - 1} \right),$$

$$\Delta p = -\frac{p_{now}}{\dfrac{1}{H_o t_{now}} - 1}, (48)$$

In (Eq. 46, 47&48), the negative value of the rate of change of momentum of light (Δp) shows that the momentum of light in T.S universe is decreased or reduced with time. The unit of (Δp) can be obtained as follows:

$$\Delta p = (kg.m/s)\left(s^{-1}.s\right) = kg.m/s,$$

Thus the unit of (Δp) is (kg.m/s).

3.2.2 The relation between (m) and (Δp):

We can find the relation between mass of photon (m) and change of momentum of photon (Δp) using (Eq. 46) as follows:

$$\Delta p = -p_{now}\left(H_o.t_{past}\right),$$

$$\Delta p = -p_{now}\left(H_o.\frac{D}{c_{past}}\right),$$

$$\Delta p = -\frac{p_{now}}{c_{past}}\left(\Delta c\right),$$

$$\Delta p = -\frac{p_{now}(1+z)}{c_{now}}(\Delta c),$$

$$\Delta p = -m(1+z)(\Delta c),$$

$$\Delta p = -m\Delta c(1+z), (49)$$

If we try to get the previous equation using (Eq. 47), we find something strange as it is shown,

$$\Delta p = -p_{past}(H_o.t_{now}),$$

$$\Delta p = -p_{past}\left(H_o.\frac{D}{c_{now}}\right),$$

$$\Delta p = -\frac{p_{past}}{c_{now}}(\Delta c),$$

$$\Delta p = -\frac{p_{past}}{c_{past}(1+z)}(\Delta c),$$

$$\Delta p = -\frac{m.\Delta c}{1+z}, (50)$$

As we use two equivalent equations that are (Eq. 46&47) to get the relation between (m) and (Δp) as it is shown,

$$-p_{now}\left(H_{\circ}.t_{past}\right) = -p_{past}\left(H_{\circ}.t_{now}\right) = \Delta p,$$

the two results of the relation between (m) and (Δp) that are obtained in (Eq. 49&50) must be the same as it is shown,

$$-m.\Delta c\left(1+z\right) = -\frac{m.\Delta c}{1+z}, (51)$$

I solve this paradox about the previous equation in chapter five (Mass in T.S universe).....

❋ ❋ ❋

CHAPTER 4:

The fifth force (New force)

A force is defined as a push or pull that changes an object's state of motion. According to Newton, an object will only accelerate if there is net force acting upon it. Newton's second law states that,

In an inertial frame of reference, the acceleration of an object (a) by a net force (F) is directly proportional to the magnitude of this force, in the same direction as the net force, and inversely proportional to the mass (m) of the object. This law can be expressed as follows:

$$a = \frac{F}{m}, (52)$$

where,

$$F = ma,$$

$$F = m\frac{\Delta v}{t}$$

$$F = \frac{\Delta p}{t}, (53)$$

Newton's second law tell us that the rate of change of momentum of an object is equal to the instantaneous force acting on it where the rate of change of momentum of an object is directly proportional to the resultant force applied and is in the direction of the resultant force. Thus if the direction of the acceleration is known, then the direction of the net force is also known and vice verse if the direction of the net force is unknown, then the direction of acceleration is unknown.

In fact, there are four fundamental forces in nature that are gravitational force, weak nuclear force, electromagnetic force and strong nuclear force.

I suggest that there is fifth force (F5), and that force causes acceleration of particles

* * *

4.1 Mathematics of the fifth force (F5)

A ccording to chapter (3), we know that momentum of light is changed or decreased (Δp) between the observer and the observed body as it is shown in (Eq. 46, 47&48).

In this section, I show the mathematical derivation of the change of momentum of light with time or the rate of change of momentum of light that are also called the fifth force (F5).

(F5) can be given using (Eq. 53) as follows:

$$F_5 = \frac{\Delta p}{t_{now}}, (54)$$

where (F5) denotes the fifth force, (Δp) denotes the change of momentum of light and (t_{now}) denotes the current time.

Using (Eq. 46), I get

$$F_5 = \frac{H_o \cdot p_{now} \cdot t_{past}}{t_{now}} = H_o \cdot p_{now}(1+z),$$

$$F_5 = H_o \cdot p_{past}, (55)$$

I get the same previuos equation using (Eq. 47) as it is shown,

$$F_5 = \frac{H_o \cdot p_{past} \cdot t_{now}}{t_{now}} = H_o \cdot p_{past},$$

The same equation (Eq. 55) can be derived using (Eq. 48) as follows:

$$F_5 = \frac{\Delta p}{t_{now}} = \left(\frac{p_{now}}{\dfrac{1}{H_o t_{now}} - 1} \right)\left(\frac{1}{t_{now}} \right),$$

$$F_5 = \frac{p_{now}}{\dfrac{1}{H_o} - t_{now}}, (56)$$

-The unit of (F_5) can be obtained as follows:

$$F_5 = H_\circ . p_{past} = s^{-1} \left(kg.m.s^{-1} \right) = kg.m.s^{-2},$$

I find that force doesn't cause acceleration of photons only, rather it accelerates all particles in T.S universe,

$$F_5 = \frac{\Delta p}{t_{now}},$$

where (Δp) denotes the change of momentum of particle.

As the change of momentum is negative (Eq. 46, 47 & 48), I find that:

"The fifth force has a negative value".

❋ ❋ ❋

4.2 The fifth force (F_5) has a constant relativistic value:

A s the light is accelerate with a constant value, I find that the change of momentum of light with time (or the value of the fifth force) must be constant also and by this way, energy is conserved.

The constant value of the fifth force can be changed only with motion where its value is decreased in moving spaces from perspective of observers at rest due to relativity of time thus "the fifth force is relativistic between inertial frames".

Fifth force: it is a negative force that causes acceleration of particles, it refers to the rate of change of momentum of the particle and equals momentum of the particle in past multiplied by Hubble's constant. It has a constant relativistic value.

Acceleration of light (a_L)	The fifth force (F_5)
Relative	Relative

If (F_5') denotes the fifth force in the moving frame with respect to the moving observer while (F_5) denotes the fifth force in that moving frame with respect to the stationary frame of reference, the relation between (F_5') & (F_5) is given as:

$$F_5 = \frac{\Delta p}{t_{now}},$$

And as time is dilated with motion with respect to the observer at rest,

$$t_{now} = \gamma t'_{now},$$

I get,

$$F_5 = \frac{\Delta p}{\gamma t'_{now}},$$

As the change of momentum of particles (Δp) is absolute or the momentum of particles is decreased equally in all frames of reference, I find that (F_5) must be given as,

$$F_5 = \frac{F'_5}{\gamma},(57)$$

Regarding to (Eq. 57), the fifth force is relative regarding to motion where it is decreased in the moving space with respect to the stationary frame of reference.

We can find the relation between the fifth force (F_5) and accceleration of light (a_l) using (Eq. 56) as follows:

$$F_5 = \frac{P_{now}}{\dfrac{1}{H_o} - t_{now}},$$

$$F_s = \frac{m.c_{now}}{\dfrac{1}{H_o} - t_{now}},$$

$$F_s = \frac{m}{\dfrac{1}{H_o.c_{now}} - \dfrac{t_{now}}{c_{now}}},$$

$$F_s = \frac{m}{\dfrac{1 - H_o.t_{now}}{H_o.c_{now}}},$$

$$F_s = \frac{m}{\dfrac{1 - H_o.t_{now}}{a_l}},$$

And as,

$$1 + z = \frac{1}{1 - H_o.t_{now}},$$

I get,

$$F_s = \frac{m}{\dfrac{1}{a_l(1+z)}},$$

So,

$$F_s = m.a_l(1+z), \quad (58)$$

* * *

4.3 The Relation between the fifth force (F₅) and wavelength (λ):

In this section, I show the mathematical relation between wavelength (λ) and the force that causes acceleration of particles (F₅).

It can be given using (Eq. 55), as follows:

$$F_5 = H_\circ . p_{past},$$

$$F_5 = H_\circ . \frac{h}{\lambda_{past}},$$

$$F_5 = \frac{H_\circ . h}{\lambda_{past}}, (59)$$

or by using (Eq. 56) as follows,

$$F_5 = \frac{P_{now}}{\dfrac{1}{H_o} - t_{now}},$$

with following the same steps in the previous section, we get:

$$F_5 = m.a_l \left(1+z\right),$$

And as,

$$E = mc^2 = h\upsilon,$$

I get,

$$F_5 = \frac{h\upsilon.a_l}{c_{now}^2}\left(1+z\right),$$

$$F_5 = \frac{H_o.h\upsilon}{c_{now}}\left(1+z\right),$$

$$F_5 = \frac{H_o.h\upsilon}{c_{now}}\left(\frac{1}{1 - H_o.t_{now}}\right),$$

$$F_5 = \frac{H_o.h\upsilon}{c_{now} - \left(H_o.c_{now}.t_{now}\right)},$$

$$F_5 = \frac{H_o.h\upsilon}{c_{now} - \left(H_o.D\right)},$$

$$F_5 = \frac{H_\circ . h\upsilon}{c_{now} - \Delta c},$$

$$F_5 = \frac{H_\circ . h\upsilon}{c_{past}},$$

$$F_5 = \frac{H_\circ . h}{\lambda_{past}},$$

The previous equation is the (Eq. 59).

We can re-write (Eq. 59) in another form as follows:

$$F_5 = \frac{H_\circ . h}{\lambda_{past}},$$

$$F_5 = \frac{a_l . h}{c_{now} . \lambda_{past}},$$

If,

$$a_l . h = Y,$$

we obtain,

$$F_5 = \frac{Y}{c_{now} . \lambda_{past}}, (60)$$

where the factor (Y) is constant, it is equal to:

$$Y = \left(6.902138691 \times 10^{-10} m.s^{-2}\right)\left(6.62607004 \times 10^{-34} kg.m^2 s^{-1}\right)$$

$$Y = 4.5734054392 \times 10^{-43} kg.m^3 / s^3$$

Regarding to (Eq. 60), the factor (Y) and the fifth force are constant while the speed of light is increased and wavelength of light in past is decreased,

$$F_5 = \frac{Y}{\uparrow c_{now} \cdot \downarrow \lambda_{past}},$$

* * *

4.4 The relation between the fifth force (F₅) and energy of photons (E)

T he relation between the fifth force (F_5) and energy of light (E) can be derived using (Eq. 56) as follows:

$$F_5 = \frac{p_{now}}{\dfrac{1}{H_o} - t_{now}},$$

$$F_5 = \frac{E}{c_{now}\left(\dfrac{1}{H_o} - t_{now}\right)},$$

$$F_5 = \frac{E}{\dfrac{c_{now}}{H_o} - D},$$

$$E = F_5\left(\frac{c_{now}}{H_o} - D\right), (61)$$

$$E = F_5\left(\frac{c_{now}^2}{a_1} - D\right), (62)$$

We can find the relation between energy (E) and the fifth force (F_5) by using (Eq. 55) as follows:

$$F_5 = H_o \cdot p_{past},$$

$$F_5 = \frac{H_o \cdot E}{C_{past}},$$

$$E = \frac{F_5 \cdot c_{past}}{H_o},$$

$$E = \frac{F_5 \cdot c_{now}}{H_o(1+z)}, (63)$$

If the two equations (Eq. 61 & 63) are valid, it must be,

$$\frac{F_5 \cdot c_{now}}{H_o(1+z)} = F_5\left(\frac{c_{now}}{H_o} - D\right),$$

$$\frac{c_{now}}{H_o(1+z)} = \frac{c_{now}}{H_o} - D,$$

$$\frac{c_{now}}{H_o(1+z)} = c_{now}\left(\frac{1}{H_o} - t_{now}\right),$$

$$\frac{1}{H_o(1+z)} = \frac{1}{H_o} - t_{now},$$

$$\frac{1}{H_o(1+z)} = \frac{p_{now}}{F_5},$$

$$F_5 = H_o \cdot p_{now}(1+z),$$

$$F_5 = H_o \cdot p_{past},$$

Thus the two equations (Eq. 61 & 63) give the same value.

Returning to (Eq. 61) I find that,

$$E = F_5\left(\frac{c_{now}}{H_o} - D\right),$$

$$E = F_5(c_{now} \cdot A_o - D),$$

$$E = F_5(D_{universe} - D),$$

$$E = F_5 \cdot \Delta D, (64)$$

where ($D_{universe}$) denotes the the diameter of the observable T.S universe , (D) denotes the distance that the light travels while (A_o) denotes the age of the universe.

✳ ✳ ✳

CHAPTER 5:

Mass in T.S universe

5.1 Mass of bodies is decreased with time

To solve (Eq. 51) in chapter (3), we must consider that masses of particles and bodies are decreased with time in T.S universe, and thus I re-formulate (Eq. 49) as follows:

$$\Delta p = -\frac{p_{now}(1+z)}{c_{now}}(\Delta c),$$

$$\Delta p = -m_{now}.\Delta c (1+z),(65)$$

and re-formulate (Eq. 50) as follows:

$$\Delta p = -\frac{p_{past}}{c_{past}(1+z)}(\Delta c),$$

$$\Delta p = -\frac{m_{past}.\Delta c}{1+z},(66)$$

where (m_{now}) denotes the current mass of the particle while (m_{past}) denotes its mass in the past. The relation between (m_{now}) & (m_{past}) can be given using (Equ. 51) as follows:

$$\Delta p = -m_{now}.\Delta c (1+z) = -\frac{m_{past}.\Delta c}{1+z},$$

$$m_{now}(1+z) = \frac{m_{past}}{1+z},$$

$$m_{now} = \frac{m_{past}}{(1+z)^2}, (67)$$

This equation shows that mass of particle is decreased with time where,

"The more progress in time, the lighter the particle will be".

* Another method to determine the relation between the current mass of particle and its mass in the past:

As the energy (E) is conserved and speed of light is increased with time, the mass of particles and bodies must be decreased with time as follows:

$$E_{now} = E_{past} = E,$$

$$m_{now} \cdot c_{now}{}^2 = m_{past} \cdot c_{past}{}^2,$$

$$m_{now} \cdot c_{past}{}^2 (1+z)^2 = m_{past} \cdot c_{past}{}^2,$$

$$m_{now} (1+z)^2 = m_{past},$$

$$m_{now} = \frac{m_{past}}{(1+z)^2},$$

* * *

5.2 The change of mass (Δm)

 s the change of speed of light (Δc) is absolute and energy is conserved, the change of mass (Δm) must be absolute also,

I find

"Mass of particles is decreased or reduced equally everywhere".

or in other words

"The change of mass (Δm) is absolute".

The change of speed of light (Δc)	The change of mass (Δm)
Absolute	Absolute

Mathematical derivation of the change of mass (Δm)

The mathematical derivation of the change of mass between past and present (Δm) can be given as,

$$\Delta m = m_{now} - m_{past}, (68)$$

$$\Delta m = m_{now} - \left(m_{now} (1+z)^2 \right),$$

$$\Delta m = m_{now} \left(1 - (1+z)^2 \right),$$

$$\Delta m = m_{now} \left(1 - \left(1^2 + 2z + z^2 \right) \right),$$

$$\Delta m = m_{now} \left(1 - 1^2 - 2z - z^2 \right),$$

$$\Delta m = m_{now} \left(-2z - z^2 \right),$$

$$\Delta m = -m_{now} \left(2z + z^2 \right),$$

$$\Delta m = -m_{now} . z (2 + z), (69)$$

And as,

$$\Delta p = -p_{now} . z = -\left(m_{now} . c_{now} \right) . z,$$

I find,

$$\frac{\Delta p}{c_{now}} = -m_{now} . z,$$

Thus (Eq. 69) can be re-formulate as,

$$\Delta m = -\frac{\Delta p}{c_{now}} (2 + z), (70)$$

The negative value of the change of mass (Δm) shows that mass of particles or bodies is decreased with time.

Regarding to (Eq. 69 & 70), it is clear that the unit of (Δm) is the kilogram where the rate of contraction of time (z) is unitless.

5.3 Rate of reduction of mass of bodies

5.3.1 Mathematics of (m_r)

We know that mass is changed or reduced (Δm) as it is shown in (Eq. 69 & 70). In this section, I show the mathematical derivation of the change of mass with time or the rate of reduction of mass.

It can be given as follows:

$$m_r = \frac{\Delta m}{t_{now}}, (71)$$

where (m_r) denotes the change of mass with time or the rate of reduction of mass.

The unit of (m_r) can be obtained as,

$$m_r = \frac{kg}{s} = kg/s,$$

5.3.2 The rate of reduction of mass has a constant relativistic value:

As the light is accelerate with a constant value, I find that the change of mass of particles with time (or the rate of reduction of mass of particles) must be constant also and by this way, energy is conserved.

The constant value of the rate of reduction of mass in T.S universe can be changed only with motion where its value in moving spaces is decreased with respect to observers at rest (stationary frame of reference) due to relativity of time thus

"The rate of reduction of mass is relativistic between inertial frames".

Acceleration of light (a_L)	The rate of reduction of mass (m_r)
Relative	Relative

If (m_r') denotes the rate of reduction of mass in a moving frame with respect to the moving observer while (m_r) denotes the rate of reduction of mass in a moving frame with respect to the stationary frame of reference, the relation between (m_r') & (m_r) is given as:

$$m_r = \frac{\Delta m}{t_{now}},$$

And as time is dilated with motion with repsect to the observer at rest,

$$t_{now} = \gamma . t'_{now},$$

I get,

$$m_r = \frac{\Delta m}{\gamma . t'_{now}},$$

As the change of mass (Δm) is absolute or the mass of particles is reduced equally in all frames of reference, I find that (m_r) must be given as,

$$m_r = \frac{m'_r}{\gamma}, (72)$$

Regarding to (Eq. 72), the rate of change of mass is relative regarding to motion where it is decreased with respect to the stationary frame of reference.

* * *

CHAPTER 6:

Gravity in T.S universe

6.1 Gravitational factor (G) is increased with time

A s the speed of light is increased with time, I find that the universal gravitational constant (G) must be increased with time in T.S universe where (G) is directly proportional to (c), as it is shown,

$$G \propto c,$$

So, I find that naming of (G) as "Gravitaional factor" is better and more accurate than "Gravitational constant".

6.1.1 Mathematics of the rate of change of gravitational factor (G_a):

If we consider that (ΔG) denotes the change of gravitational factor and (G_a) denotes the change of gravitational factor with time or the rate of change of gravitational factor, the relation between (ΔG) and (G_a) can be determined as follows:

$$G_a = \frac{\Delta G}{t_{now}}, (73)$$

Lets consider that time (t_{now}) is equal to the age of the universe (A_o), the change of gravitational factor (ΔG) in this case is equal to the current gravitational factor (G_{now}) as the gravitational factor at birth of the universe ($G_{initial}$) is equal to zero, as it is shown:

$$G_a = \frac{\Delta G}{A_o},$$

$$G_a = \frac{G_{now} - G_{initial}}{A_o},$$

$$G_{initial} = zero,$$

$$G_a = \frac{G_{now}}{A_o},$$

And as,

$$A_o = \frac{1}{H_o},$$

I get,

$$G_a = H_o.G_{now}, (74)$$

The unit of the change of gravitational factor with time (G_a) is given as,

$$G_a = s^{-1}\left(m^3.kg^{-1}.s^{-2}\right),$$

$$G_a = m^3.kg^{-1}.s^{-3},$$

6.1.2 Calculation the value of (G_a):

Using the previous equation (Eq. 74), we can calculate the value of the change of gravitational factor with time or the rate of change of gravitational factor in T.S universe that is constant as follows,

$$G_a = H_o.G_{now},$$

$$G_a = \left(2.3007 \times 10^{-18}\right)\left(6.67407898 \times 10^{-11}\right),$$

$$G_a = 1.5355053509286 \times 10^{-28}\, m^3 . kg^{-1} . s^{-3},$$

6.1.3 The rate of change of gravitational factor (Ga) has a constant relativistic value:

As the light accelerating with a constant value, I find that the change of gravitational factor with time (Ga) must be constant also.

The constant value of the rate of change of gravitational factor (Ga) can be changed only with motion where it is decreased in moving spaces with respect to the observer at rest due to relativity of time between inertial frames thus "the rate of gravitational factor is relativistic between inertial frames".

If (Ga′) denotes the rate of change of gravitational factor in a moving frame with respect to the moving observer while (Ga) denotes the rate of change of gravitational factor in that moving frame with respect to the stationary frame of reference, the relation between (Ga′) & (Ga) is given as:

$$G_a = \frac{\Delta G}{t_{now}},$$

And as time is dilated with motion with respect to the   observer at rest,

$$t_{now} = \gamma . t'_{now},$$

I get,

$$G_a = \frac{\Delta G}{\gamma . t'_{now}},$$

As the change of gravitational factor (ΔG) is absolute or the gravitational factor is increased equally in all frames of reference, I find that (G_a) must be given as,

$$G_a = \frac{G'_a}{\gamma} , (75)$$

Regarding to (Eq. 75), the rate of change of gravitational factor (G_a) is relative with motion where it is decreased with respect to the stationary frame of reference.

6.2 Gravitational factor (G) in past and present

I f (Gnow) denotes the current gravitational factor while (Gpast) denotes the gravitational factor in the past, the relation between (Gnow) & (Gpast) can be given as follows:

$$G_a = \frac{\Delta G}{t_{now}},$$

$$\Delta G = G_a . t_{now},$$

Using (Eq. 74), I find:

$$\Delta G = \left(H_\circ . G_{now}\right) . t_{now}, (76)$$

$$\Delta G = \left(H_\circ . G_{now}\right) \frac{D}{c_{now}},$$

$$\Delta G = H_\circ . D \left(\frac{G_{now}}{c_{now}}\right),$$

$$\Delta G = \Delta c \left(\frac{G_{now}}{c_{now}} \right),$$

$$\frac{\Delta G}{\Delta c} = \frac{G_{now}}{c_{now}},$$

And as,

$$\frac{\Delta G}{\Delta c} = \left(\frac{\Delta G}{\Delta c} \right)_{now} = \left(\frac{\Delta G}{\Delta c} \right)_{past},$$

I get,

$$\frac{G_{now}}{c_{now}} = \frac{G_{past}}{c_{past}},$$

$$\frac{G_{now}}{c_{past}\left(1+z\right)} = \frac{G_{past}}{c_{past}},$$

$$G_{now} = G_{past}\left(1+z\right), (77)$$

where,

"The universal gravitational factor (G) is increased equally every-where".
"The change of gravitional factor (ΔG) is absolute".
"The rate of change of gravitional factor (Ga) has a constant relativ-istic value".

Problems:

Dr Mohamed Abdelwhab Husseiny

Problem one:
What is the value of gravitational factor (G) after one century?
- (Gnow)=6.67408×10-11m3.kg-1.s-2.

Solution
-As the gravitational factor is increased with time, the current value of (Gnow) that is equal to (6.67408×10-11 m3kg-1s-2) will be increased to be equal (Gfuture), and as the relation between (Gnow) & (Gpast) is given as,

$$G_{now} = G_{past}\left(1+z\right),$$

the relation between (Gnow) & (Gfuture) can be given as,

$$G_{future} = G_{now}\left(1+z\right),$$

And as,

$$1+z = \frac{1}{1-H_o.t_{future}},$$

I get,

$$G_{future} = \frac{G_{now}}{1-H_o.t_{future}},$$

We must don't forget that time in the static universe is contracted, the presence of word "after" in "after one century" in this problem means that time (one century) is related to the future's perspective or the observer's perspective,

$$t_{future} = 100\,years = 100\times365.2422\times24\times60\times60,$$

$$t_{future} = 3155692608\,sec,$$

-By the previous knowledge, we can solve this problem

114

$$G_{future} = \frac{G_{now}}{1 - H_o . t_{future}},$$

$$G_{future} = \frac{6.67408 \times 10^{-11}}{1 - \left(\left(2.3007 \times 10^{-18} \right) \left(3155692608 \right) \right)},$$

$$G_{future} = \frac{6.67408 \times 10^{-11}}{1 - \left(7.2580929984 \times 10^{-9} \right)},$$

$$G_{future} = 6.674080048 \times 10^{-11} m^3 . kg^{-1} . s^{-2},$$

This is the value of the gravitational factor (G) after (one century).

Problem two:
What is the value of gravitational factor 21 centuries ago?
-(Gnow)=6.67408×10-11m3.kg-1.s-2.

Solution
-The presence of word "ago" in "21 centuries ago" in this problem means that time (21 centuries) is related to the past perspective, so:

$$t_{past} = 21 \times 100 \times 365.2422 \times 24 \times 60 \times 60,$$

$$t_{past} = 66269544768 \text{ sec},$$

Thus we can solve this problem as follows:

$$G_{now} = G_{past} \left(1 + z \right),$$

$$G_{now} = G_{past}\left(1+H_{\circ}.t_{past}\right),$$

$$G_{past} = \frac{G_{now}}{1+H_{\circ}.t_{past}},$$

$$G_{past} = \frac{6.67408\times10^{-11}}{1+\left(\left(2.3007\times10^{-18}\right)\left(66269544768\right)\right)},$$

$$G_{past} = 6.67407898273719\times10^{-11}m^{3}.kg^{-1}.s^{-2},$$

This is the value of the gravitational factor (G) since 21 centuries.

❖ ❖ ❖

6.3 The relation between the change of gravitational factor (ΔG) and the change of speed of light (Δc)

I n this section, I derive the mathematical relation between the change of gravitational factor (ΔG) and that of the speed of light (Δc). Using the equation in the previous section that,

$$\frac{\Delta G}{\Delta c} = \frac{G_{now}}{c_{now}},$$

I find,

$$\Delta G = \left(\frac{G_{now}}{c_{now}} \right) \Delta c,$$

$$\Delta G = \left(\frac{6.67408 \times 10^{-11}}{3 \times 10^{8}} \right) \Delta c,$$

$$\Delta G = \left(2.225 \times 10^{-19} \right) \Delta c,$$

And if,

$$2.225 \times 10^{-19} = Q,$$

We obtain,

$$\Delta G = Q . \Delta c, (78)$$

where the value (Q) is constant. The unit of (Q) can be given as,

$$Q = \frac{\Delta G}{\Delta c},$$

$$Q = \frac{m^3 . kg^{-1} . s^{-2}}{m . s^{-1}} = m^2 . kg^{-1} . s^{-1},$$

Thus,

$$Q = 2.225 \times 10^{-19} \ m^2 / kg . s,$$

* * *

6.4 Relation between the change of gravitational factor (ΔG) and distance (D)

I n this section, I derive the mathematical relation between the distance (D) and the difference in the gravitational factor (ΔG) (the change of gravitational factor) using (Eq. 76) as follows:

$$\Delta G = \left(H_o.G_{now} \right) t_{now},$$

$$\Delta G = \left(H_o.G_{now} \right) \frac{D}{c_{now}},$$

$$\Delta G = \left(\frac{H_o.G_{now}}{c_{now}} \right) D,$$

Regarding to (Eq. 74) that is,

$$H_o.G_{now} = G_a,$$

I get,

$$\Delta G = \left(\frac{G_a}{c_{now}} \right) D,$$

$$\Delta G = \left(\frac{1.5355053509286 \times 10^{-28}}{3 \times 10^8} \right) D,$$

$$\Delta G = \left(5.118351169762 \times 10^{-37} \right) D,$$

And if,

$$5.118351169762 \times 10^{-37} = w,$$

we get,

$$\Delta G = w.D, (79)$$

where (w) is a constant,

The unit of (w) can be determined as,

$$w = \frac{\Delta G}{D} = \frac{m^3 .kg^{-1}.s^{-2}}{m} = m^2 .kg^{-1}.s^{-2},$$

*Problems

Problem one:
What is the difference in gravitational factor (ΔG) between two galaxies separated by 7×1022 m?

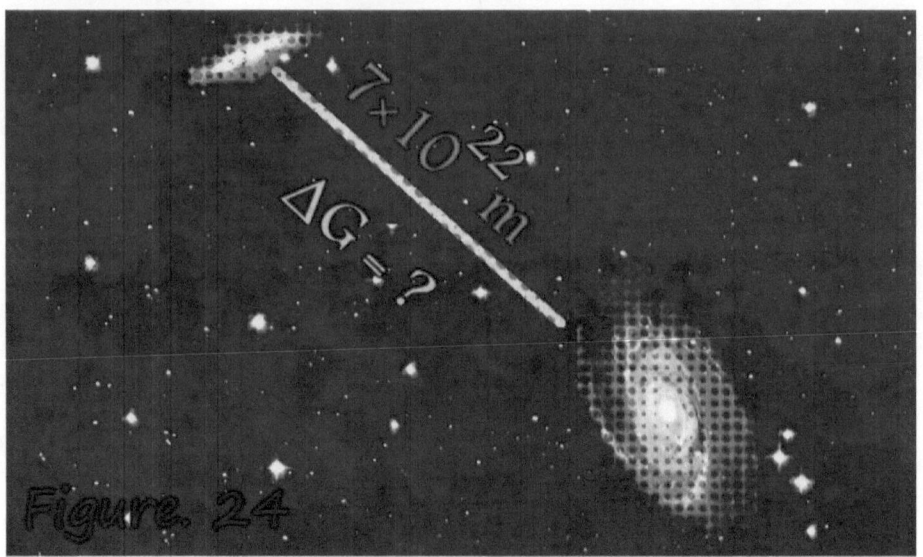

Figure 24

<u>Solution</u>

$$\Delta G = w.D,$$
$$\Delta G = \left(5.118351169762 \times 10^{-37}\right)\left(7 \times 10^{22}\right),$$
$$\Delta G = 35.828 \times 10^{-15} m^3 .kg^{-1} .s^{-2},$$

that is the value of the difference in (G) between the two galaxies.

✳ ✳ ✳

6.5 Relation between the change of gravitational factor (ΔG) and the orbit's radius (r)

R egarding to the previous section, the relation between the change of gravitational factor ΔG) and the distance (D) is given as,

$$\Delta G = w.D,$$

where (w) is a constant factor. We know that the relation between the distance (D) & radius of the circle or the distance from the center to the orbit (r) is given as,

$$D = 2r,$$

So, the relation between the change of gravitational factor (ΔG) and the distance from the center to the orbit (r) can be given as,

$$\Delta G = 2w.r, (80)$$

To make (Eq. 80) a general equation, I re-write in another form as,

$$\Delta G = 2w.\Delta r, (81)$$

where (Δr) in (Eq. 81) denotes the distance from the center to an orbit or the distance from an orbit to another one (Fig. 25).

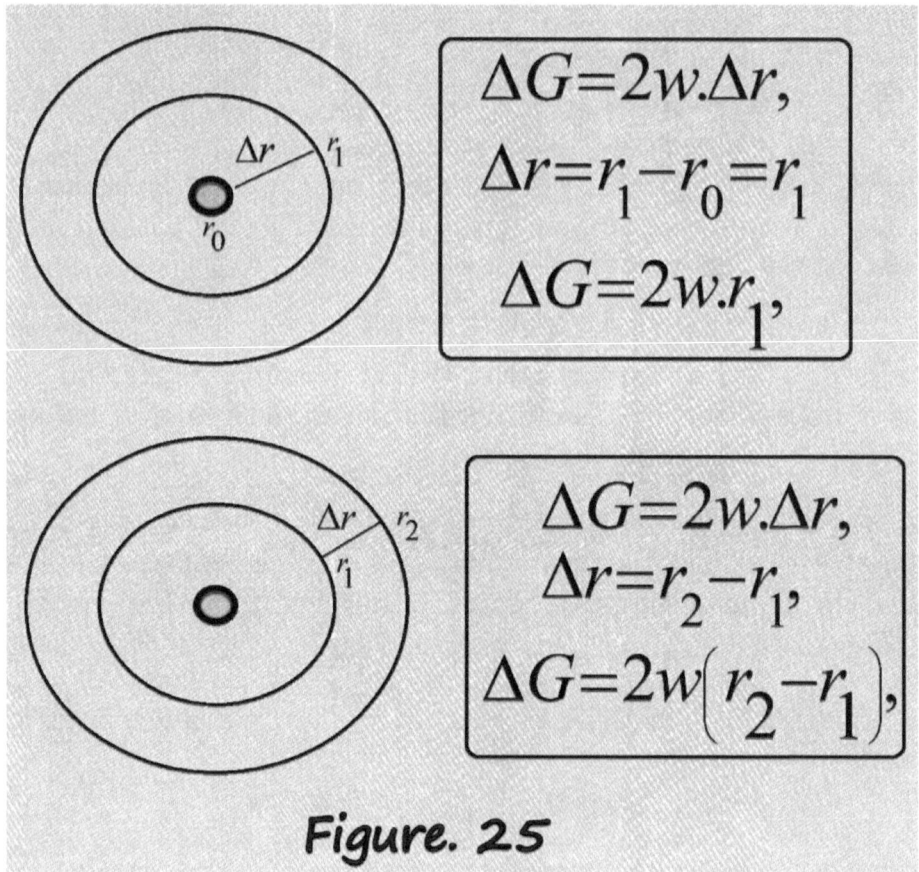

$$\Delta G = 2w.\Delta r,$$
$$\Delta r = r_1 - r_0 = r_1$$
$$\Delta G = 2w.r_1,$$

$$\Delta G = 2w.\Delta r,$$
$$\Delta r = r_2 - r_1,$$
$$\Delta G = 2w\left(r_2 - r_1\right),$$

Figure. 25

*Problems

Problem one:
What is the difference in gravitational factor or the change of gravitational factor (ΔG) between two bodies (a) and (b) where the first body (a) exists at distance (6×106 m) from the center of the orbit while the other body (b) exists at distance (3×109 m)?

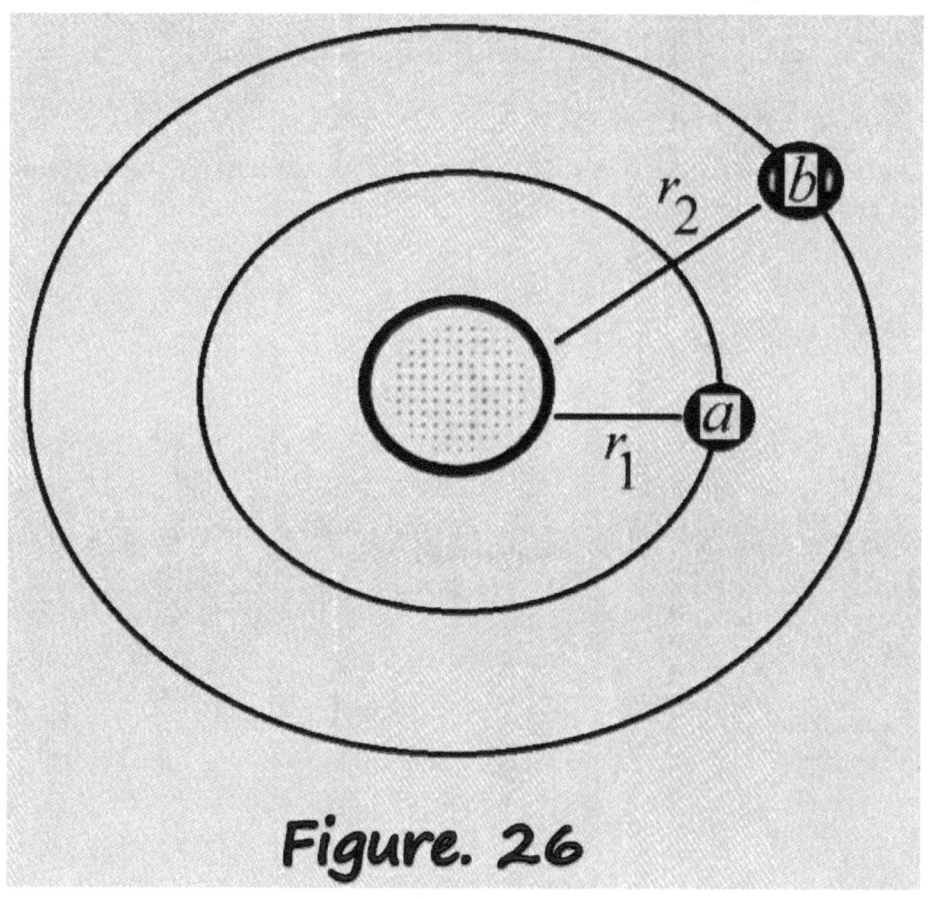

Figure. 26

<u>Solution</u>

To solve this problem, we use (Eq. 81) that

$$\Delta G = 2w.\Delta r,$$

Thus,

$$\Delta G = 2w\left(\left(3\times10^9\right)-\left(6\times10^6\right)\right),$$

$$\Delta G = 2w\left(2994\times10^6\right),$$

$$\Delta G = 2\left(5.118351169762\times10^{-37}\right)\left(2994\times10^6\right),$$

$$\Delta G = \left(2\times5.118351169762\times2994\right)\times10^{-31},$$

$$\Delta G = 30648.686804534856 \times 10^{-31},$$

$$\Delta G = 3.0648686804534856 \times 10^{-27} \, m^3 . kg^{-1} . s^{-2},$$

that is the value of the difference in gravitational factor between the two bodies.

* * *

6.6 Orbital velocity of stars (v) is independent the distance from the center of the galaxy

The third Kepler's law shows that the square of orbital period of a planet is proportional to the cube of the semi major axis of its orbit, or in other words, the larger axis of orbit, the smaller the orbital speed and vice verse. Kepler's third law shows that velocity (v) is inversely proportional to the distance from the center of the orbit (r). This is because of the gravitational force being exerted on the planets by the space, as it is shown:

$$v = \sqrt{\frac{G.M}{r}}, (82)$$

Where (M) denotes the mass of the center of the orbit.

The change in speed of light with time makes gravitational factor (G) is changed also with time, thus we can re-write (Eq. 82) as,

$$v = \sqrt{\frac{\Delta G.M}{\Delta r}}, (83)$$

where (ΔG) denotes the difference in gravitational factor while (Δr) denotes the difference in the orbit radius.
Accordingly,

$$v^2 = \frac{\Delta G.M}{\Delta r},$$

$$\Delta G = \frac{v^2}{M}.\Delta r, (84)$$

From (Eq. 81 & 84) I get,

$$2w = \frac{v^2}{M},$$

$$w = \frac{v^2}{2M}, (85)$$

where the orbital velocity of the body (v) is given as,

$$v = \sqrt{2w.M}, (86)$$

Regarding to (Eq. 86), I find that the orbital velocity of stars doesn't depend on the radius of the orbit or the distance from the center of the galaxy, rather it depends on (w) that is a constant and the mass of the center of the galaxy that is constant also.

I find that the change of gravitational factor with time leads to constancy of the orbital velocity of stars at any radius from the center of the galaxy (Fig. 27).

-(Eq. 86) can be re-written as,

$$v = \sqrt{2M\left(\frac{G_a}{c_{now}}\right)}, (87)$$

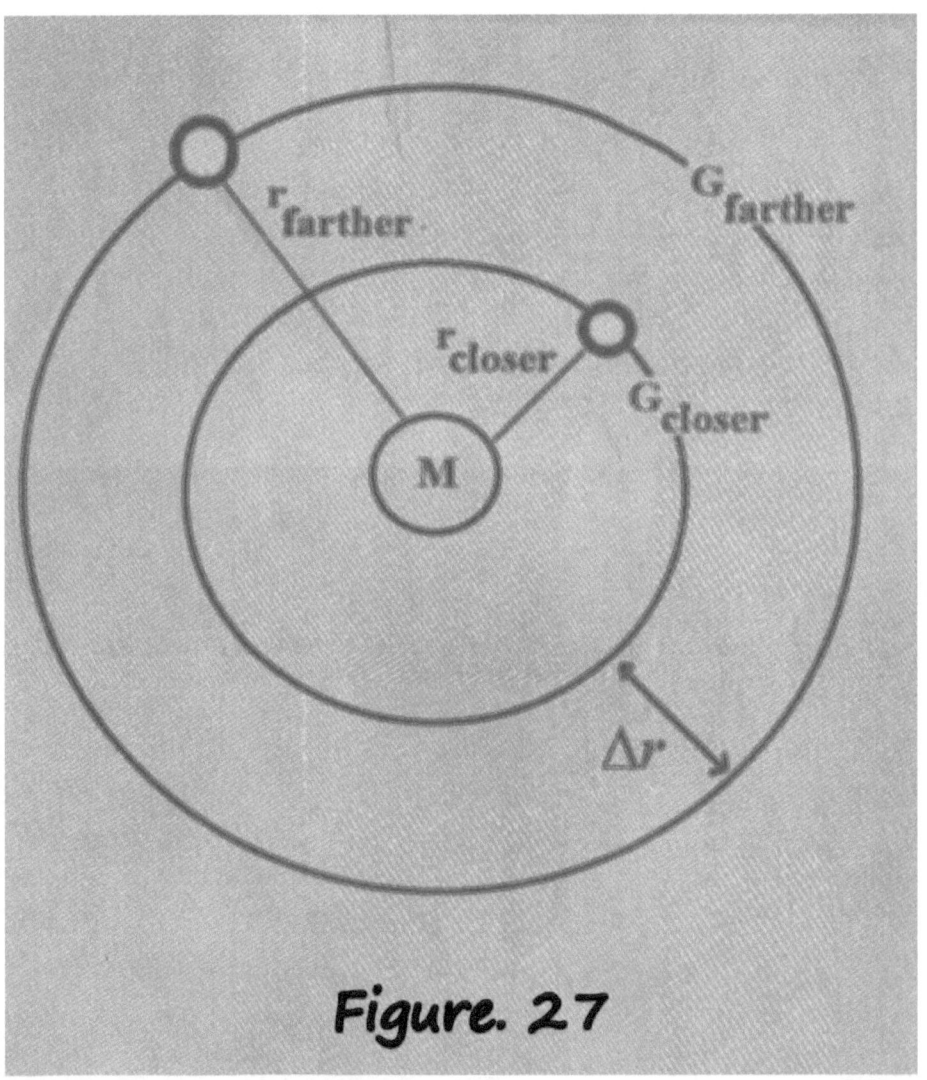

Figure. 27

* * *

CHAPTER 7:

Einstein gravitational constant "kappa" in T.S universe

7.1 Einstein's Kappa (k) is decreased with time

E instein's constant or Einstein's gravitational constant, denoted by (k) or (*kappa*), is the coupling constant appearing in the Einstein's field equation, it is determined as,

$$k = \frac{8\pi G}{c^4}, (88)$$

I find that constant (k) must be changed with time because speed of light (c) and gravitational factor (G) are changed with time. I called (k) as "Einstein's factor", I find that the relation between the current Einstein's factor (k_{now}) and its value in past (k_{past}) can be given as follows:

$$\left(8\pi\right)_{now} \left(8\pi\right)_{past},$$

$$\frac{k_{now}.c_{now}^{4}}{G_{now}} = \frac{k_{past}.c_{past}^{4}}{G_{past}},$$

$$\frac{k_{now}.c_{past}^{4}\left(1+z\right)^{4}}{G_{past}\left(1+z\right)} = \frac{k_{past}.c_{past}^{4}}{G_{past}},$$

$$\frac{k_{now}.\left(1+z\right)^{4}}{\left(1+z\right)} = k_{past},$$

$$k_{now}.\left(1+z\right)^{3} = k_{past},$$

$$k_{now} = \frac{k_{past}}{(1+z)^3}, (89)$$

From (Eq. 89), it is clear that Einstein's factor (k) is decreased with time.

7.2 The change of Einstein's factor (Δk)

A s the change of speed of light (Δc) and that of gravitational factor (ΔG) are absolute, the change of Einstein's factor (Δk) must be absolute also.

$$\Delta k = \frac{8\pi . \Delta G}{\Delta c^4},$$

"Einstein's factor (k) is changed or decreased equally in all frames of reference".

or in other words

"The change of Einstein's kappa (k) is absolute".

The change of speed of light (Δc)	The change of gravitational factor(ΔG)	The change of Einstein's factor (Δk)
Absolute	Absolute	Absolute

*Mathematical derivation of the change of Einsten's factor (Δk)
The mathematical derivation of the change of Einstein's factor between past and present (Δk) can be given as,

$$\Delta k = k_{now} - k_{past},$$

$$\Delta k = k_{now} - \left(k_{now} \left(1+z\right)^3 \right),$$

$$\Delta k = k_{now} \left(1 - \left(1+z\right)^3 \right), (90)$$

It is clear that (Δk) in (Eq. 90) equals a negative value. The negative value of the rate of change of Einstein's kappa (Δk) in the temporary static universe shows that Einstein's factor is decreased with time. The unit of (Δk) is (s2m-1kg-1) or (N-1).

* * *

7.3 Rate of change of Einstein's factor (k_a)

7.3.1 Mathematics of (k_a)

We know that Einstein's factor is changed or decreased with time as it is shown in (Eq. 89). In this section, I show the mathematical derivation of the change of Einstein's factor with time or the rate of change of Einstein's factor (k_a).

$$k_a = \frac{\Delta k}{t_{now}}, (91)$$

$$k_a = \frac{k_{now}\left(1-(1+z)^3\right)}{t_{now}}, (92)$$

The unit of the rate of change of Einstein's factor (k_a) is given as,

$$k_a = \frac{s^2.m^{-1}.kg^{-1}}{s},$$

$$k_a = s.m^{-1}.kg^{-1},$$

7.3.2 The rate of change of Einstein's factor has a constant relativistic value

As the rate of change of gravitational factor (G_a) and the rate of change of speed of light (acceleration) (a_l) are constant, I find that the rate of change of Einstein factor must be constant also. The constant value of the rate of change of Einstein's factor (k_a) in T.S universe can be changed only with motion where, its value in moving spaces is decreased with respect to observers at rest (stationary frame of reference) due to relativity of time thus "The rate of change of Einstein's factor is relativistic between inertial frames".

Acceleration of light (a_l)	The rate of change of gravitational factor (G_a)	The rate of change of Einstein's factor (k_a)
Relative	Relative	Relative

If (k_a') denotes the rate of change of Einstein's factor in a moving frame with respect to the moving observer while (k_a) denotes the rate of change of Einstein's factor in that moving frame with respect to the stationary frame of reference, the relation between (k_a') & (k_a) is given as:

$$k_a = \frac{\Delta k}{t_{now}},$$

And as time is dilated with motion with respect to the observer at rest,

$$t_{now} = \gamma t'_{now},$$

I get,

$$k_a = \frac{\Delta k}{\gamma t'_{now}},$$

As the change of Einstein's factor (Δk) is absolute or Einstein's factor is decreased equally in all frames of reference, I find that (k_a) must be given as,

$$k_a = \frac{k'_a}{\gamma}, (93)$$

Regarding to (Eq. 93), the rate of change of Einstain's factor is relative regarding to motion where it is decreased with respect to the stationary frame of reference.

* * *

CHAPTER 8:

Cosmic microwave background radiation (CMB)
of T.S universe

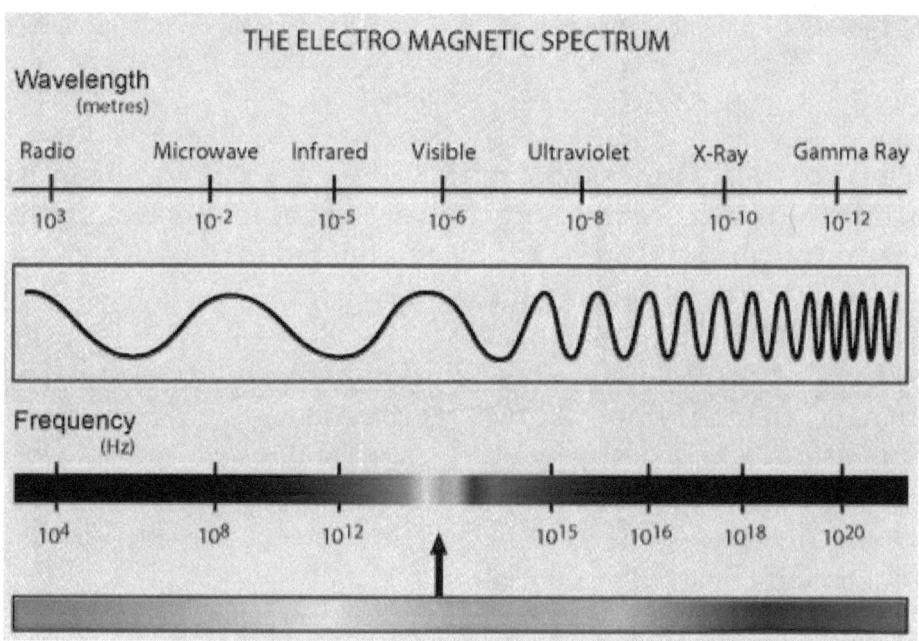

8.1 Temperature of T.S universe is decreased with time

ll objects emit electromagnetic radiation, and the amount of radiation emitted at each wavelength depends on the temperature of the object .

Wien's law, states that,

"the black body radiation curve for different temperature peaks at a wavelength is inversely proportional to the temperature".

$$\lambda \propto \frac{1}{T},$$

Thus hot objects emit more of their light at short wavelengths while cold objects emit more of their light at long wavelengths where the temperature of an object is related to the wavelength at which the object gives out the most light.

According to Wien's law, the relation between wavelength of light (λ) and the temperature (T) is given as,

$$\lambda.T = b, (94)$$

where (b) is a constant of proportionality called Wien's displacement constant and equals,

$$b = 2.8977729 \times 10^{-3} m.k,$$

The oldest light we can see, the farthest back both in time and space that we can look. This light set out on its journey more than 14 billion years ago, long before the earth or even our galaxy existed. As the energy of the oldest light is conserved and speed of that light is increased with time in T.S universe, the wavelength of the oldest light goes to be stretched with time to be enters into the microwave part of the electromagnetic spectrum and thus the temperature of T.S universe is decreased with time.

The oldest light is called the cosmic microwave background radiation (CMB). It is a form of electromagnetic radiation discovered in 1965 that fills the entire universe. It is an emission of uniform, black body thermal energy coming from all parts of the sky.

If (λ_{now}) denotes the current wavelength of CMB and (T_{now}) denotes the current temperature of CMB, while (λ_{past}) denotes the wavelength of CMB in past and (T_{past}) denotes the temperature of CMB in the past, the relation between (T_{now}) & (T_{past}) can be given using Wien's law as follows:

$$b_{now} = b_{past} = b,$$

$$\lambda_{now}.T_{now} = \lambda_{past}.T_{past},$$

$$\left(\lambda_{past}\left(1+z\right)\right)T_{now} = \lambda_{past}.T_{past},$$

$$\left(1+z\right)T_{now} = T_{past},$$

$$T_{now} = \frac{T_{past}}{1+z},(95)$$

Thus, the temperature of the cosmic microwave background radiation (CMB) or the temperature of T.S universe is decreased with time.

$$T_{now} = T_{past}\left(1 - H_{o}.t_{now}\right),(96)$$

＊ ＊ ＊

8.2 The change of temperature of T.S universe (ΔT)

As the wavelength of CMB is stretched equally in all frames of reference, I find that temperature of CMB or temperature of T.S universe must has the same behaviour where,

"Temperature of T.S universe is decreased equally in all frames of reference".

or in other words

"The change of temperature of T.S universe (ΔT) is absolute".

The change of wavelength of CMB ($\Delta\lambda$)	The change of temperature of CMB (ΔT)
Absolute	Absolute

In this section, I show the mathematical derivation of the change of temperature of T.S universe (ΔT) as follows,

$$\Delta T = T_{now} - T_{past},$$
$$\Delta T = T_{now} - \left(T_{now}\left(1+z\right)\right),$$
$$\Delta T = T_{now}\left(1-\left(1+z\right)\right),$$
$$\Delta T = T_{now}\left(1-1-z\right),$$
$$\Delta T = -T_{now}.z, (97)$$

Using (Eq. 39) that,

$$z = \cfrac{1}{\left(\cfrac{1}{H_o.t_{now}}\right)-1},$$

I get,

$$\Delta T = -\cfrac{T_{now}}{\left(\cfrac{1}{H_o.t_{now}}\right)-1}, (98)$$

The negative value of the change of temperature of T.S universe (ΔT) shows that temperature of T.S universe is decreased with time.

The (Eq. 98) can be written in another form as,

$$\Delta T = -T_{now}\left(H_{\circ}.t_{past}\right), (99)$$

The unit of (ΔT) is Kelvin (K),

* * *

8.3 The rate of change of temperature of T.S universe (T_a)

8.3.1 Mathematics of (T_a)

The mathematical derivation of the change of temperature of T.S universe with time or the rate of change of temperature of T.S universe or the cooling rate of T.S universe (T_a) can be given as follows:

$$T_a = \frac{\Delta T}{t_{now}} , (100)$$

Using (Eq. 98), I get,

$$T_a = \frac{1}{t_{now}} \times \frac{T_{now}}{\left(\dfrac{1}{H_o.t_{now}} \right) - 1},$$

$$T_a = \frac{T_{now}}{\left(\dfrac{1}{H_o} \right) - t_{now}} , (101)$$

Or by another way using (Eq. 99) as,

$$T_a = \frac{\Delta T}{t_{now}},$$

$$T_a = \frac{T_{now}\left(H_{\circ}.t_{past}\right)}{t_{now}},$$

$$T_a = T_{now}.H_{\circ}\left(1+z\right),$$

$$T_a = H_{\circ}.T_{past},\left(102\right)$$

The unit of the rate of change of temperature of T.S universe (T_a) is given as,

$$T_a = \frac{K}{s} = K/s,$$

8.3.2 The rate of change of temperature of T.S universe has a constant relativistic value

As the rate of change of wavelength of CMB (v_λ) is constant, I find that the rate of change of temperature of CMB (T_a) must be constant also.

The constant value of the rate of change of temperature of T.S universe can be changed only with motion where its value in moving spaces is decreased with respect to observers at rest (stationary frame of reference) due to relativity of time thus "The rate of change of temperature of T.S universe is relativistic between inertial frames".

The rate of change of wavelength of CMB (v_λ)	The rate of change of temperature of CMB (T_a)
Relative	Relative

If (T_a') denotes the rate of change of temperature of CMB in a moving space with respect to the moving observer while (T_a) denotes the rate of change of temperature of CMB in that moving space with respect to the stationary frame of reference, the relation between (T_a') & (T_a) is given as:

$$T_a = \frac{\Delta T}{t_{now}},$$

And as time is dilated with motion with respect to the observer at rest,

$$t_{now} = \gamma t'_{now},$$

I get,

$$T_a = \frac{\Delta T}{\gamma t'_{now}},$$

As the change of temperature of CMB (ΔT) is absolute or as the temperature of T.S universe is decreased equally in all frames of reference, I find that (T_a) must be given as,

$$T_a = \frac{T'_a}{\gamma},(103)$$

Regarding to (Eq. 103), the rate of change of temperature of CMB is relative regarding to motion where temperature of CMB of moving space is decreased with respect to that of the stationary frame of reference.

8.3.3 Calculation the value of the change of temperature of T.S universe per one sec (T_a)$_1$

Dr Mohamed Abdelwhab Husseiny

Regarding to observations, the current temperature of the cosmic microwave background radiation (T_{now}) as observed in the present day is equal to (2.725K). Thus the change of temperature of CMB per one sec (T_a)₁ can be calculated using (Eq. 101) as,

$$T_a = \frac{T_{now}}{\left(\dfrac{1}{H_o}\right) - t_{now}},$$

$$\left(T_a\right)_1 = \frac{2.725}{\left(\dfrac{1}{2.3007\times 10^{-18}}\right) - 1\,\text{sec}},$$

$$\left(T_a\right)_1 = 6.2694075\times 10^{-18}\ K/s,$$

Thus the temperature of T.S universe is decreased with (6.26×10-18) every one second.

✳ ✳ ✳

CHAPTER 9:

Calculation of important values

9.1 Calculation the value of the current wavelength of CMB radiation (λnow)

The temperature of the cosmic microwave background radiation as observed in the present day (T_{now}) is equal to (2.725K).

Dr Mohamed Abdelwhab Husseiny

Thus the current wavelength of the cosmic microwave background radiation (λ_{now}) can be calculated using Wien's law as,

$$\lambda_{now} = \frac{b}{T_{now}},$$

$$\lambda_{now} = \frac{2.8977729 \times 10^{-3} m.K}{2.725 K},$$

$$\lambda_{now} = 1.0634028990826 \times 10^{-3} m,$$

* * *

9.2 Calculation the value of speed of elongation or stretching of wavelength of CMB radiation (v_λ)

(v_λ) can be calculated using (Eq. 23) as,

$$v_\lambda = H_o . \lambda_{now},$$

$$v_\lambda = (2.3007 \times 10^{-18} s^{-1})(1.0634028990826 \times 10^{-3} m),$$

$$v_\lambda = 2.44657104991933782 \times 10^{-21} m/s,$$

146

* * *

9.3 Calculation the value of the current momentum of CMB radiation (p~now~)

(p~now~) can be calculated as,

$$p_{now} = \frac{h}{\lambda_{now}},$$

$$p_{now} = \frac{6.62607004 \times 10^{-34}}{1.0634028990826 \times 10^{-3}},$$

$$p_{now} = 6.231006183748723 \times 10^{-31} \, kg.m/s,$$

* * *

9.4 Calculation the value of the change of momentum of CMB radiation per one second or the value of the fifth force (F₅)

(F₅) can be calculated as,

$$F_5 = \frac{P_{now}}{\dfrac{1}{H_o} - t_{now}} \, ,$$

$$F_5 = \frac{6.231006183748723 \times 10^{-31}}{\dfrac{1}{2.3007 \times 10^{-18}} - 1} \, ,$$

$$F_5 = 1.433567592695 \times 10^{-48} \, kg.m/s^2 \, ,$$

❉ ❉ ❉

9.5 Calculation the value of the smallest mass in T.S universe (the current mass of CMB radiation)

(m_{now}) can be determined as,

$$p_{now} = m_{now} \cdot c_{now},$$

$$m_{now} = \frac{p_{now}}{c_{now}},$$

$$m_{now} = \frac{6.231006183748723 \times 10^{-31}}{3 \times 10^{8}},$$

$$m_{now} = 2.07700206124957 \times 10^{-39} \, kg,$$

* * *

9.6 Calculation the value of reduction of mass per one second (m_r)

As the time is contracted with 2.3007×10-18 per the second, I find that the mass of particles is reduced during the second with value that is calculated as,

$$m_r = \frac{\Delta m}{t_{now}},$$

$$m_r = \frac{m_{now}.z(2+z)}{t_{now}},$$

$$m_r = \frac{\left[\left(2.07700206124957\times10^{-39}\right)\left(2.3007\times10^{-18}\right)\right]\left[2+\left(2.3007\times10^{-18}\right)\right]}{1},$$

$$m_r = 9.5571172\times10^{-57} \ kg/s,$$

* * *

c_{now}	3×10^8 m/s
a_L	$6.902138691 \times 10^{-10}$ m/s^2
λ_{now}	$1.0634028990826 \times 10^{-3}$ m
v_λ	$2.44657104991933782 \times 10^{-21}$ m/s
p_{now}	$6.231006183748723 \times 10^{-31}$ kg.m/s
F_5	$1.433567592695 \times 10^{-48}$ N
m_{now}	$2.07700206124957 \times 10^{-39}$ kg
m_r	$9.5571172 \times 10^{-57}$ kg /s
G_{now}	$6.67407898 \times 10^{-11}$ m^3. kg^{-1}. s^{-2}
G_a	$1.5355053509286 \times 10^{-28}$ m^3. kg^{-1}. s^{-3}
T_{now}	2.725 K
T_a	$6.2694075 \times 10^{-18}$ K/s
k_{now}	1.866×10^{-26} m/kg
k_a	$1.28793186 \times 10^{-43}$ m.kg^{-1}.s^{-1}

CHAPTER 10:

Velocity, Linear momentum, mass and wave-length of bodies

According to the previous chapters, speed of light and its wavelength are increased with time while momentum of light and its mass are decreased with time, I find that results dosen't relate to photons only, rather it is related to all particles in T.S universe where,

"Velocity of particles in T.S univers is increased with time".

"Wavelength of particles in T.S universe is stretched with time".

"Linear momentum of particles in T.S universe is decreased with time".

And also

"Mass of particles in T.S universe is reduced with time".

❋ ❋ ❋

10.1 Equations related to velocity of particles:

$$v_{now} = \frac{v_{past}}{1 - H_{o}.t_{now}}, (104)$$

$$\Delta v = +H_{o}.D, (105)$$

$$v_{a} = H_{o}.v_{now} \ (106)$$

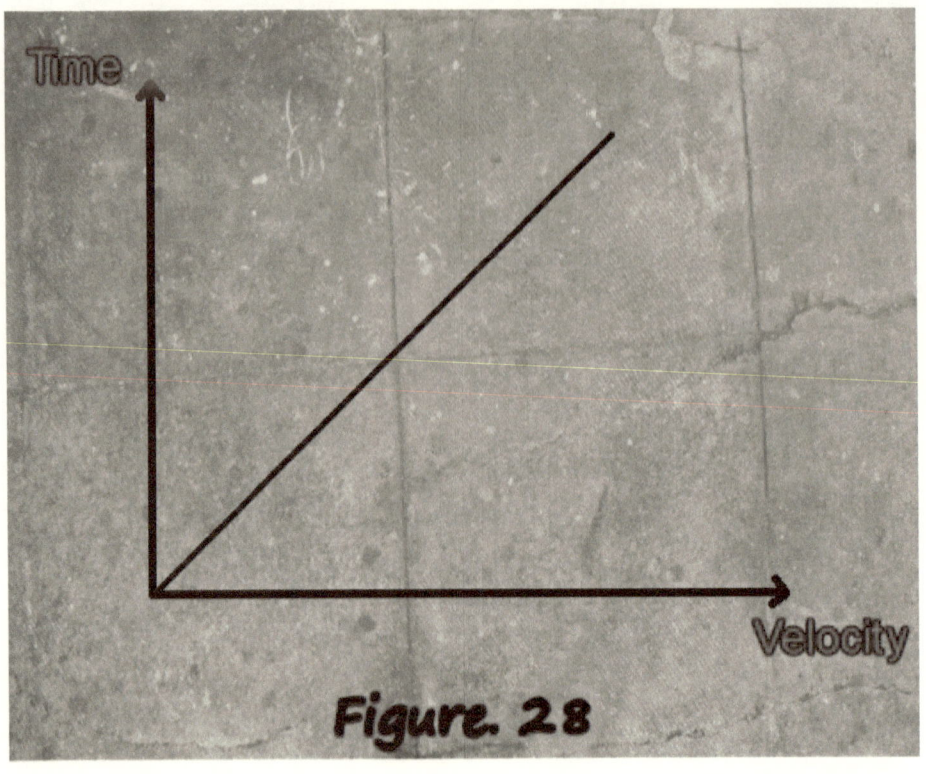

Figure. 28

* * *

10.2 Equations related to wavelength of particles:

$$\lambda_{now} = \frac{\lambda_{past}}{1 - H_{\circ}.t_{now}}, (107)$$

$$\Delta\lambda = +H_{\circ}.\lambda_{now}.t_{now}, (108)$$

$$v_{\lambda} = H_{\circ}.\lambda_{now}, (109)$$

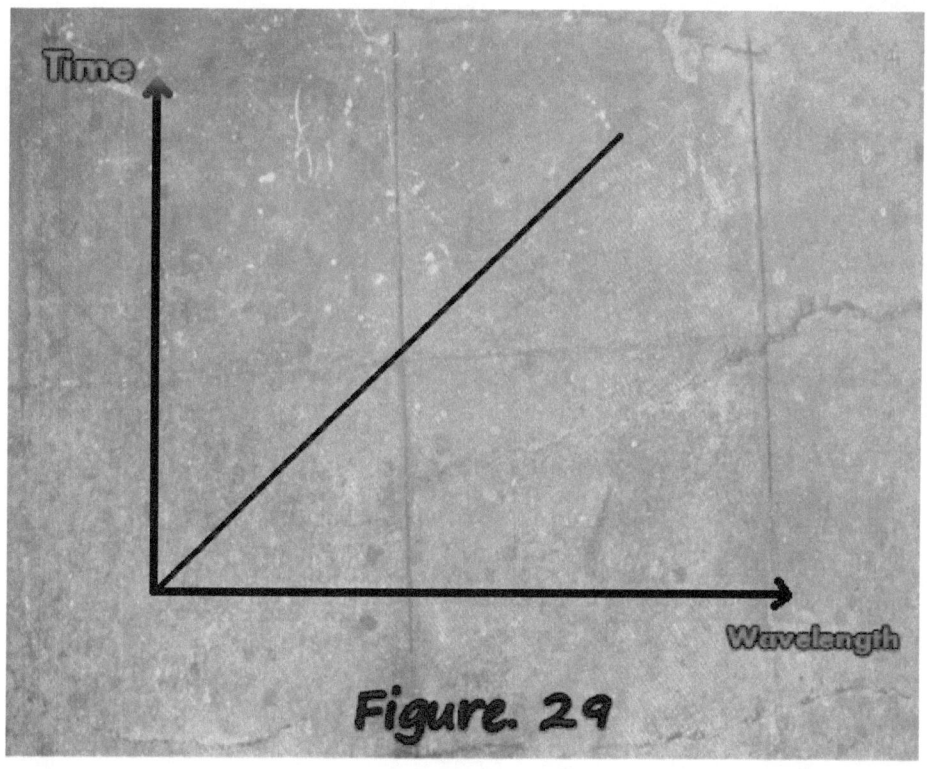

Figure. 29

* * *

10.3 Equations related to Linear momentum of particles:

$$p_{now} = p_{past}\left(1 - H_{o}.t_{now}\right), (110)$$

$$\Delta p = -\frac{p_{now}}{\dfrac{1}{H_{o}.t_{now}} - 1}, (111)$$

$$F_{s} = \frac{p_{now}}{\dfrac{1}{H_{o}} - t_{now}}, (112)$$

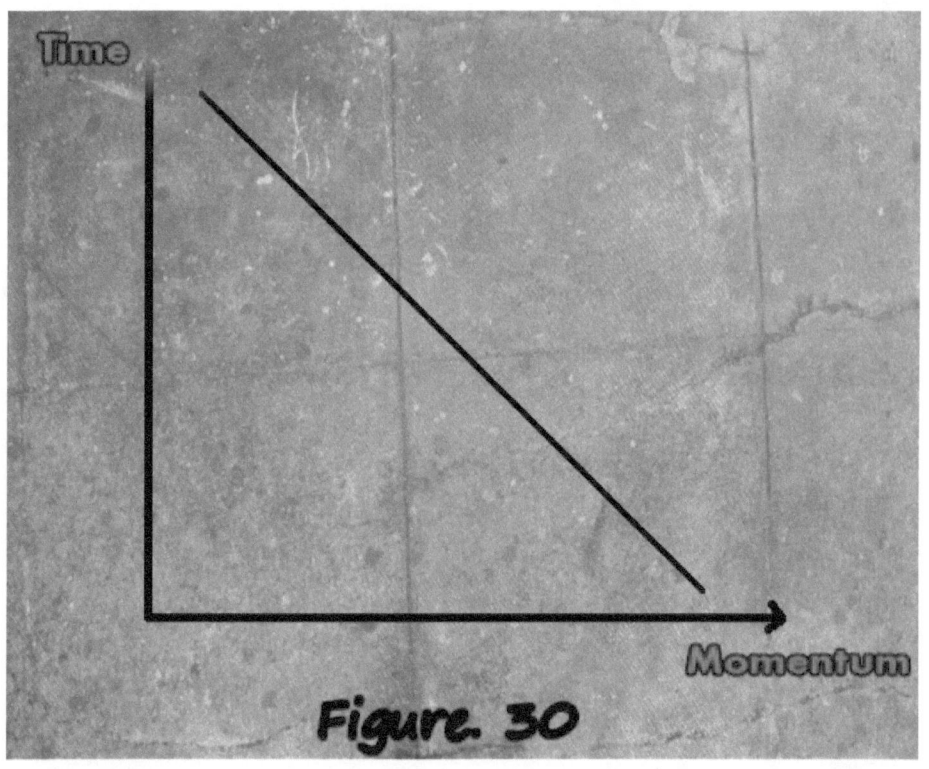

Figure. 30

* * *

10.4 Equations related to mass of particles:

$$m_{now} = m_{past} \left(1 - H_{\circ}.t_{now}\right)^2, (113)$$

$$\Delta m_{now} = -\frac{\Delta p}{c_{now}}(2+z),(114)$$

$$m_r = \frac{\Delta p}{D}(2+z),(115)$$

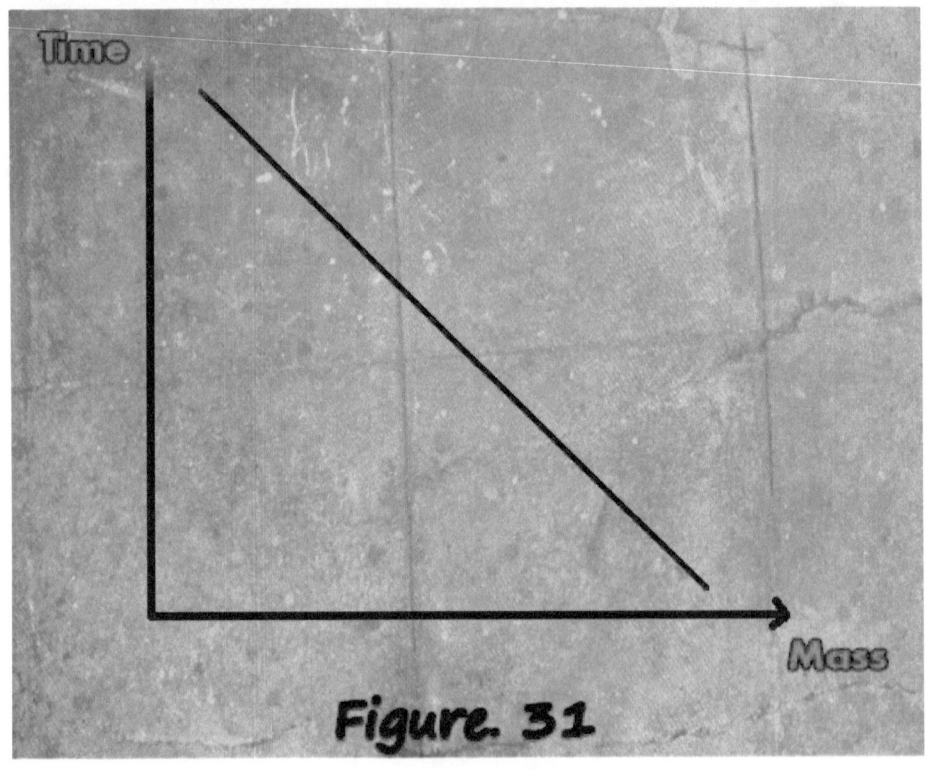

Figure. 31

* * *

PART THREE:

"Simultaneous Multiverse"

or

"Time-separated universe"

CHAPTER 1:

Speed of light in absence of observation & at observation

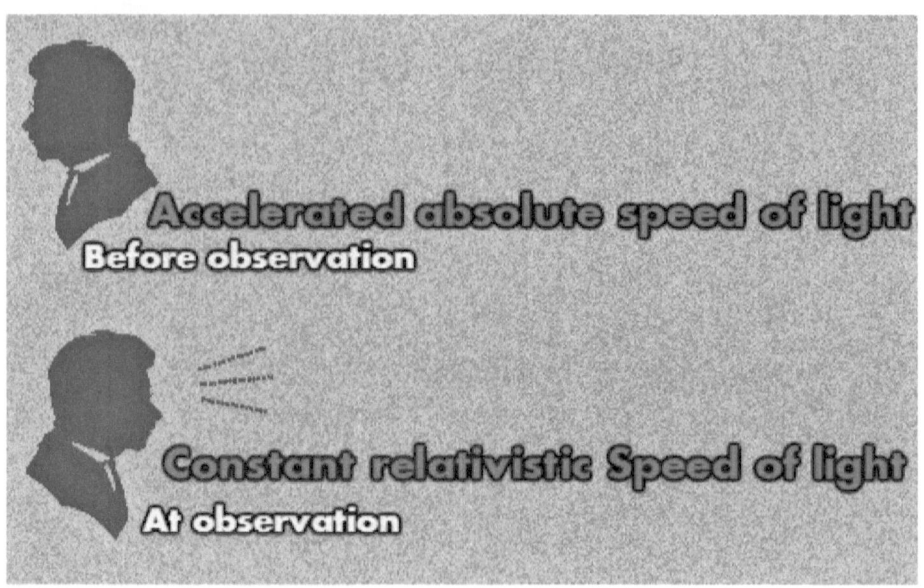

1.1 Speed of light in absence of observation (Space relativity)

A s the speed of light is increased equally everywhere, speed of light will be the same in all frames of reference in absence of observation or in darkness.
I find that, in absence of observation light has accelerated absolute speed, therefore the nature of bodies and particles in absence of observation are subjected to *"accelerated absolute speed of light"*.

In absence of observation:

1. Light has accelerated absolute speed.

2. Time of particles is absolute as the speed of light is absolute.

3. Space of particles is relative as the light is accelerated.

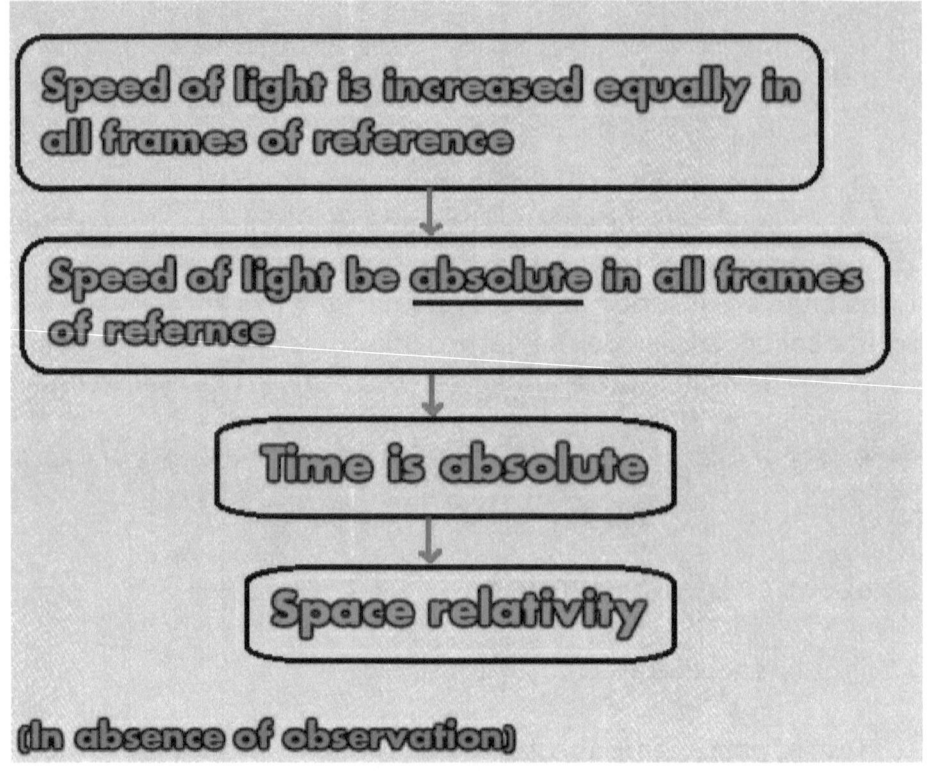

* * *

1.2 Speed of light at observation (Time relativity)

A s the speed of light is increased with time and as the process of observation requires time, I find that at observation; there is a difference in speed of light between the observer and the observed body, thus the light at observation has a relativistic speed.

For example, if time interval between the two bodies in is equal to 3 sec, the observer will see the state of the moving body 3 sec ago where the speed of light since 3 sec is equal to c1, so the difference in speed of light is appear at observation or in other words, speed of light is relative at observation.

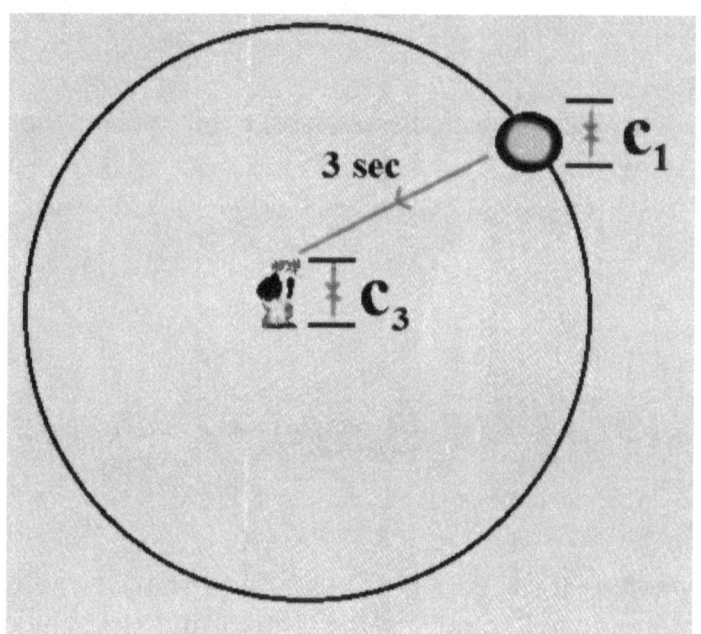

The relativity of speed of light between bodies at observation leads to relativity of time at observation where the light cone of the observed body differs from that of the observer.

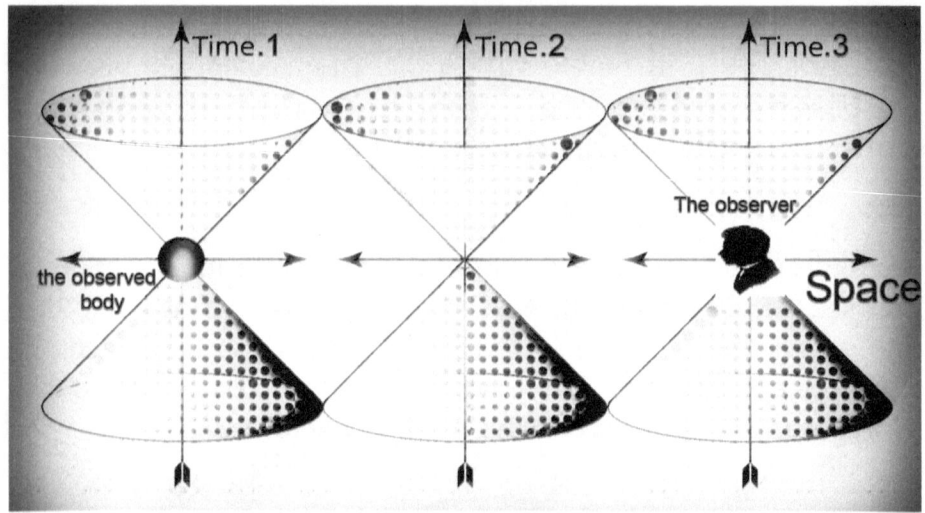

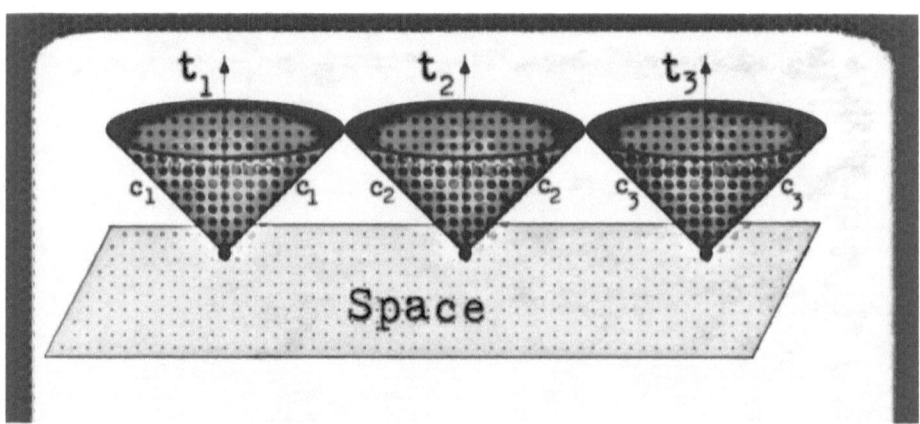

As the speed of light at the observer and that at the observed body are increased equally with time, I find that the observation process doesn't allow the phenomenon of acceleration of light to be appeared thus the relativistic speed of light be constant where it dosen't change with time with respect to the observer.

I find that light at observation has a constant relativistic speed, therefore the nature of particles (bodies) at observation are subjected to *"constant relativistic speed of light"*.

At observation:

1. Light has a constant relativistic speed.

2. Time of the observed body is relative.

3. Space of the observed body is absolute (static universe).

❈ ❈ ❈

Briefly;

<u>In absence of observation,</u>
1. Each particle has its own universe "space".
2. All particles share the same time.

At observation,

1. Each particle has its own time.
2. All particles share the same space "universe".

* * *

CHAPTER 2:

Simultaneous multiverse or Time separated universe

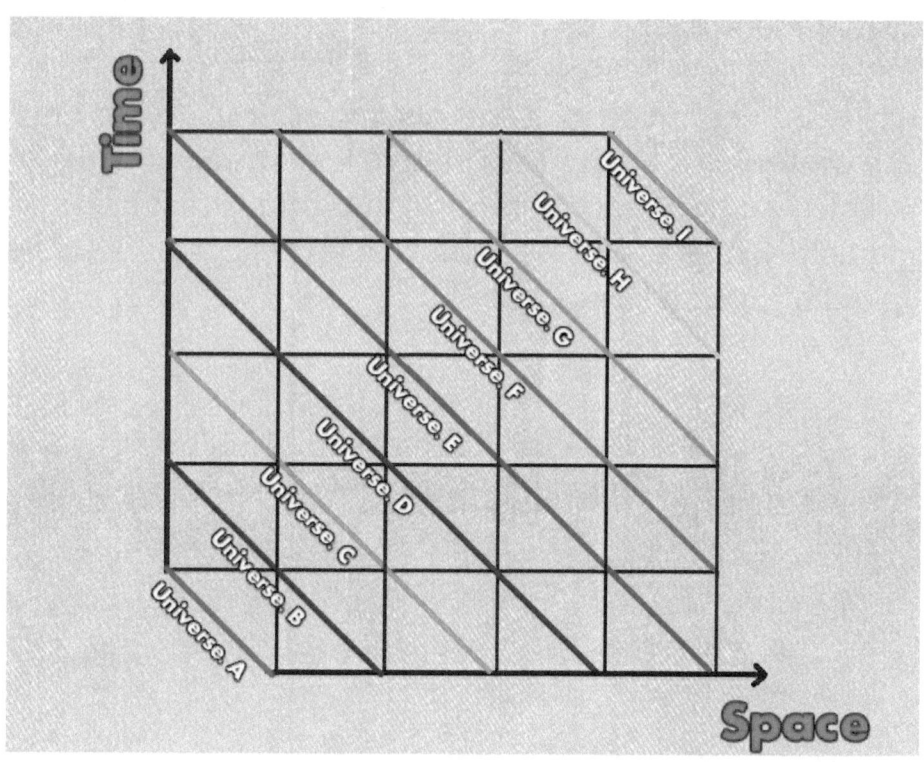

I find that space is sloped or tilted with respect to time, where

1. Before observation:

Particles are exist in **Simultaneous multiverse** where space dimension is relative between particles as the speed of light is accelerated while time is absolute as the speed of light is absolute..

2. At observation:

Particles are exist in **Time-separated universe** where the dimension of time is relative as the speed of light is relative between the observer and the observed particle while space is absolute as the acceleration of speed of light is disappeared.

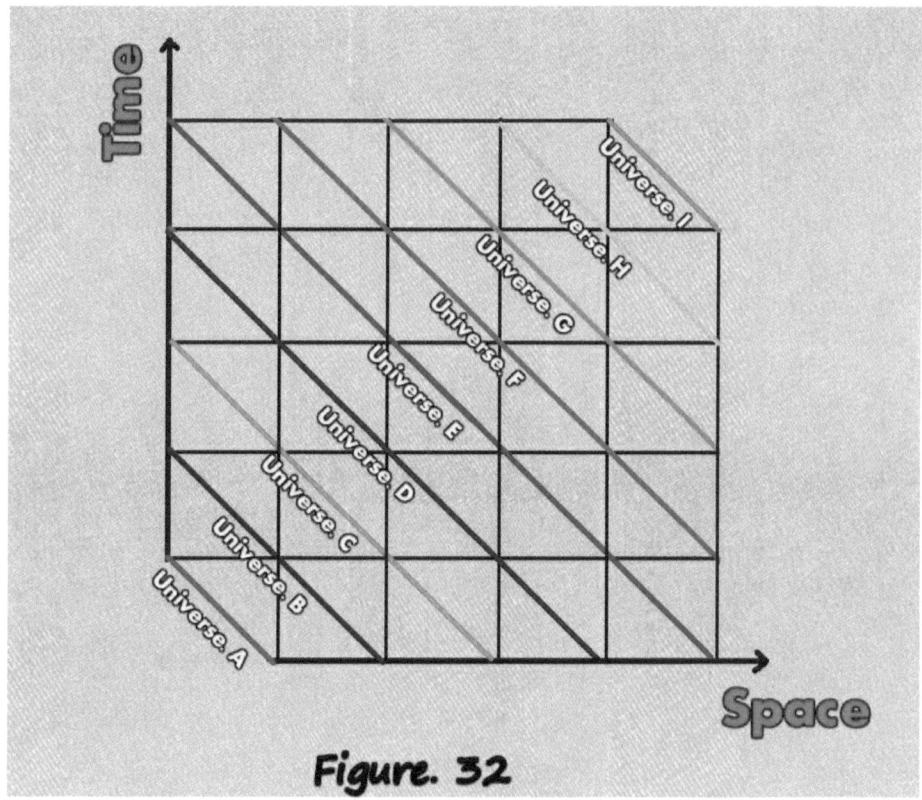

Figure. 32

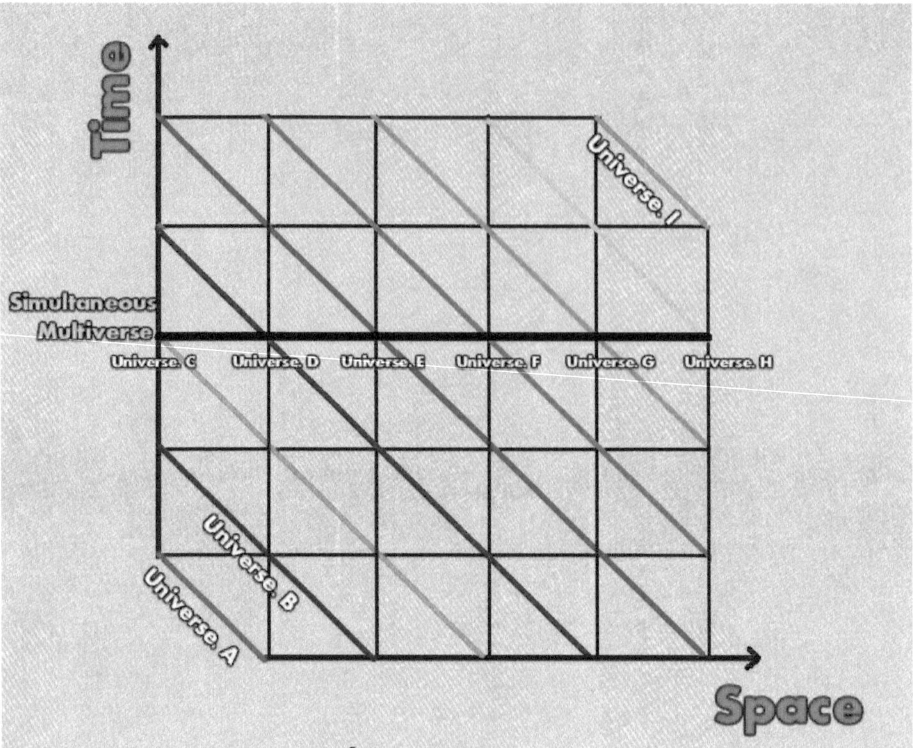

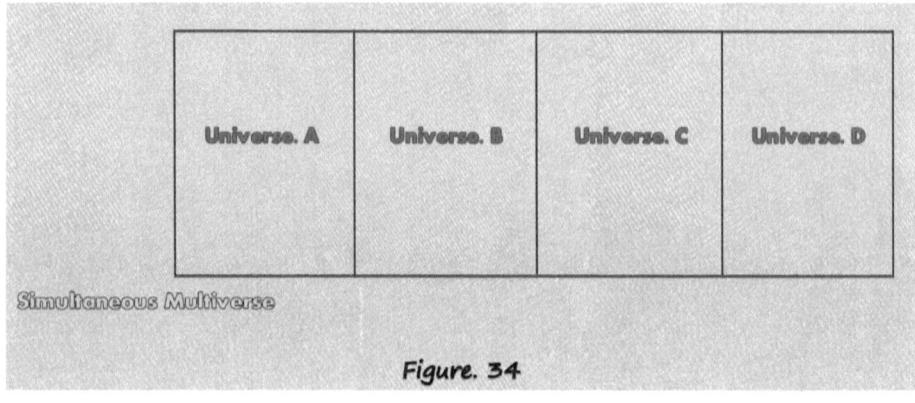

Figure. 33

Figure. 34

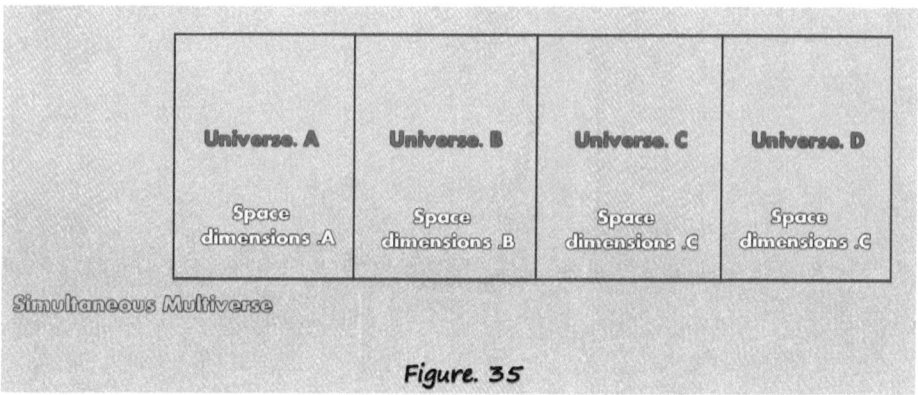

Figure. 35

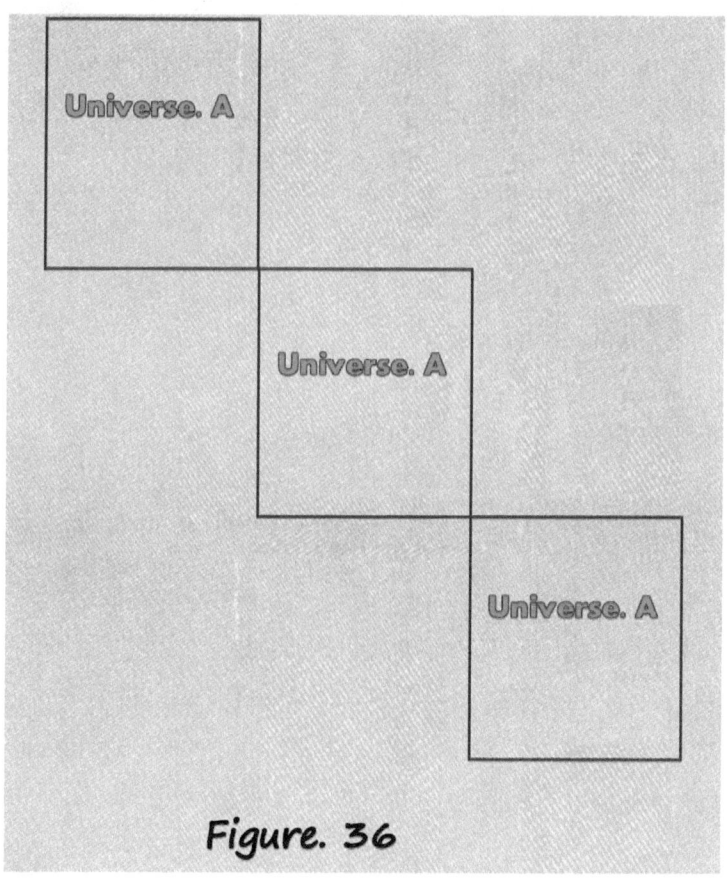

Figure. 36

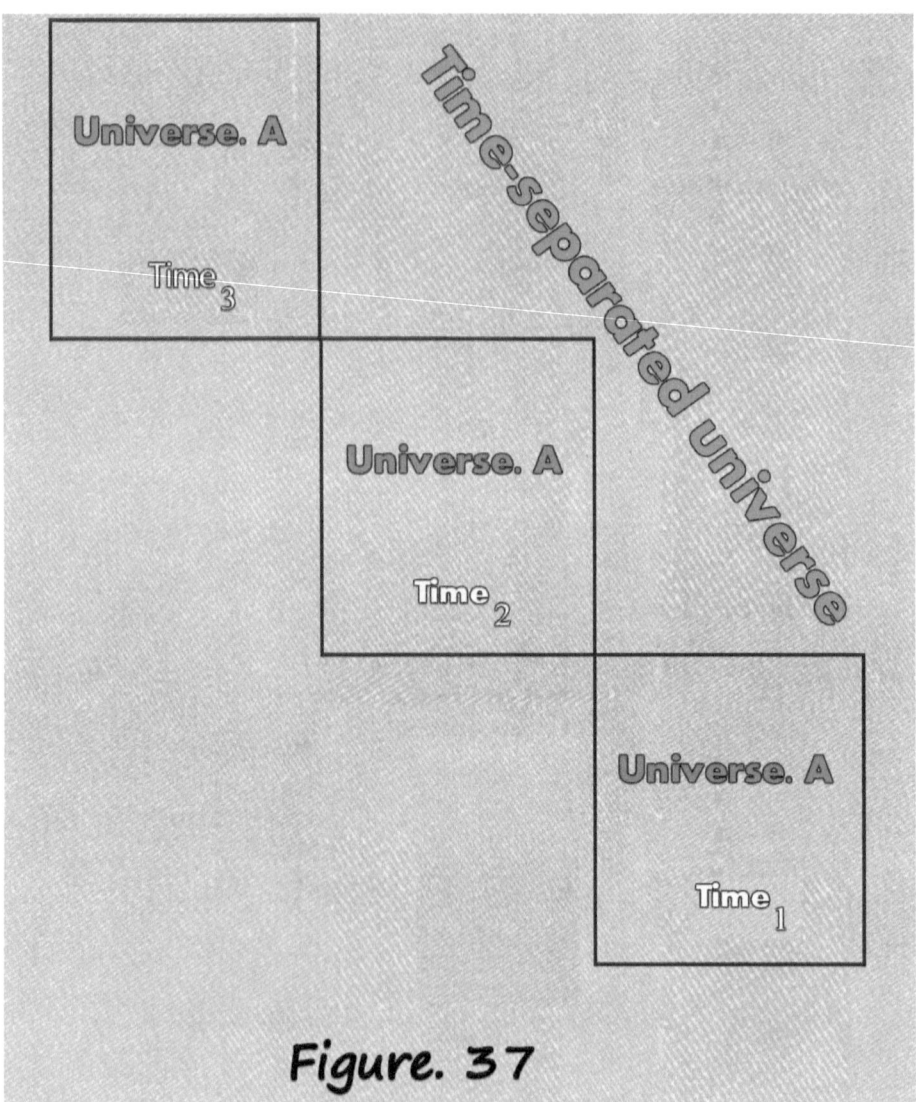

Figure. 37

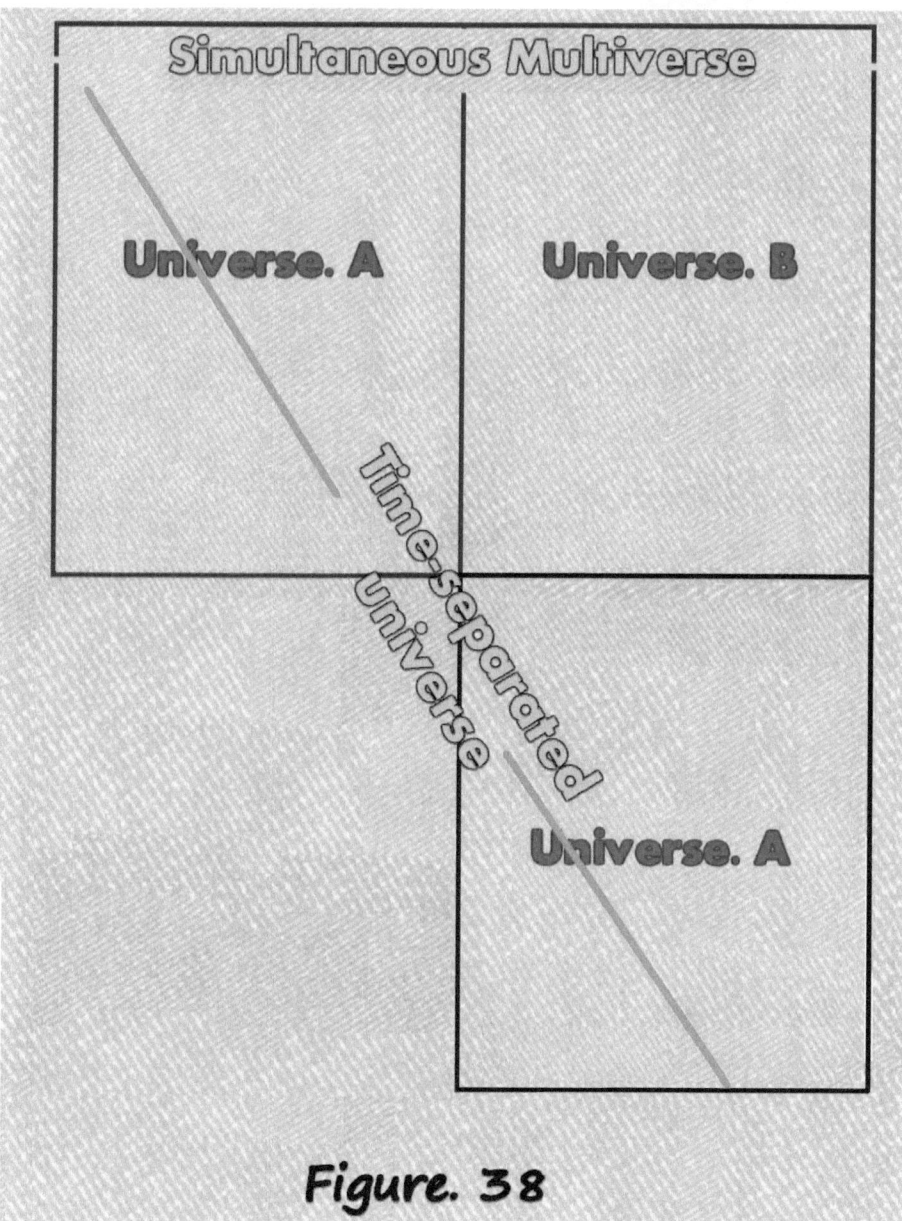

Figure. 38

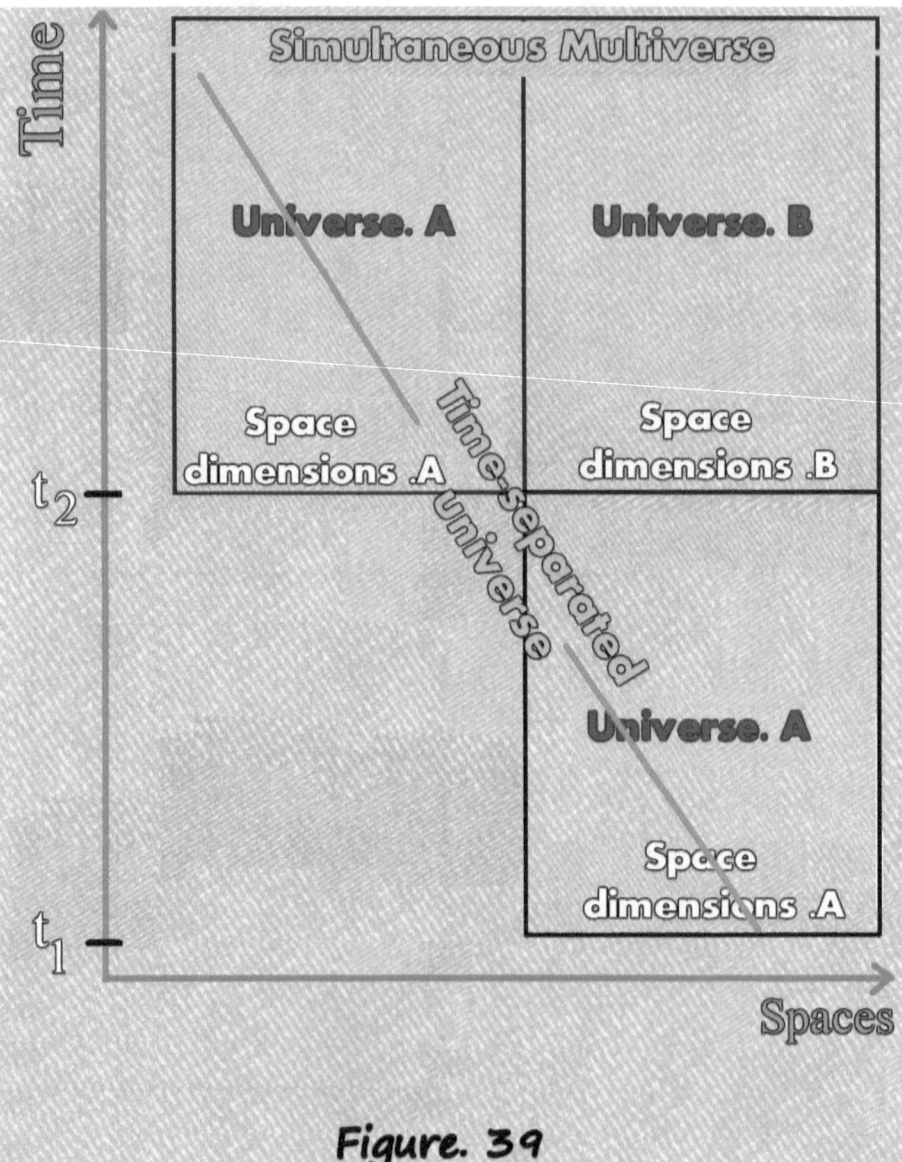

Figure. 39

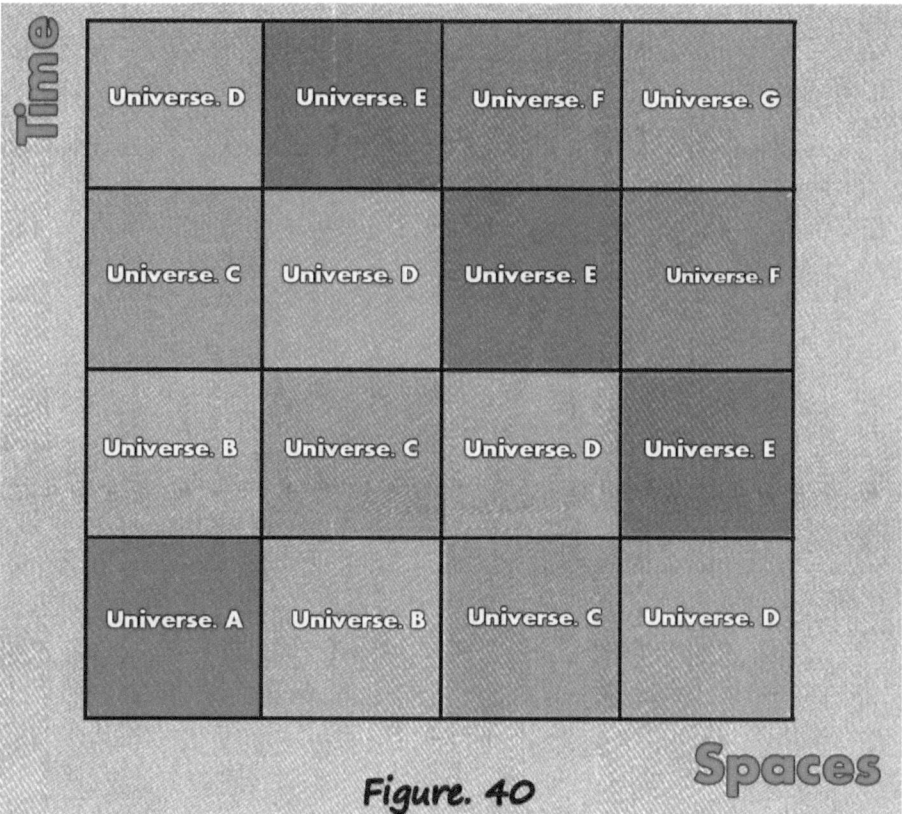

Figure. 40

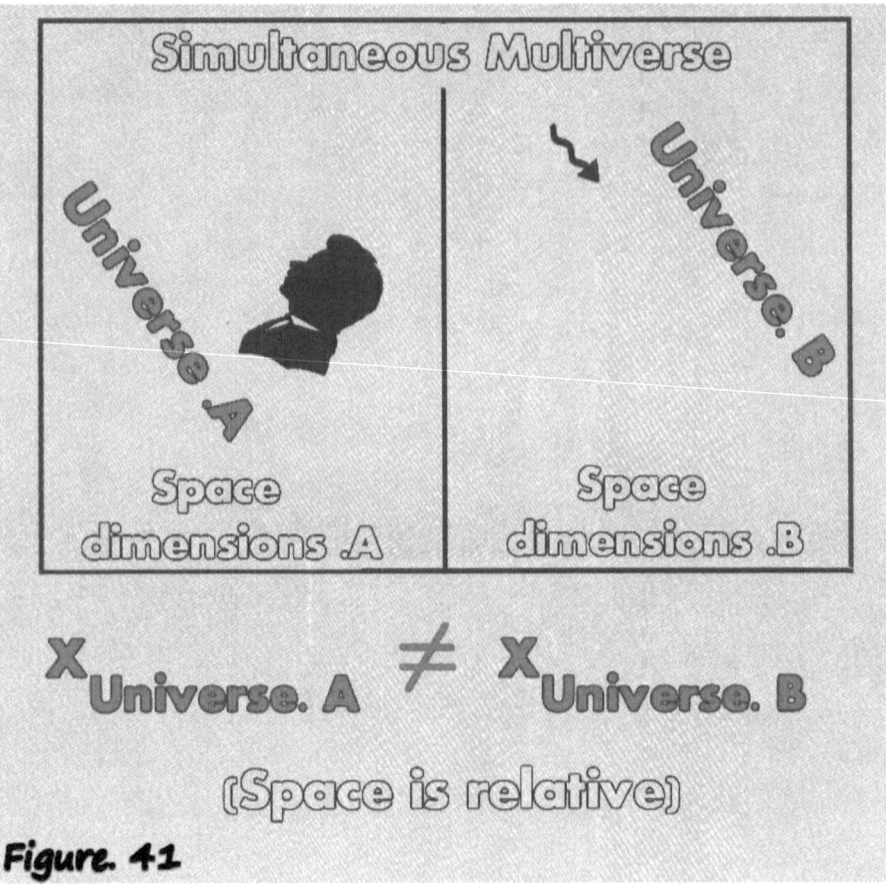

Figure. 41

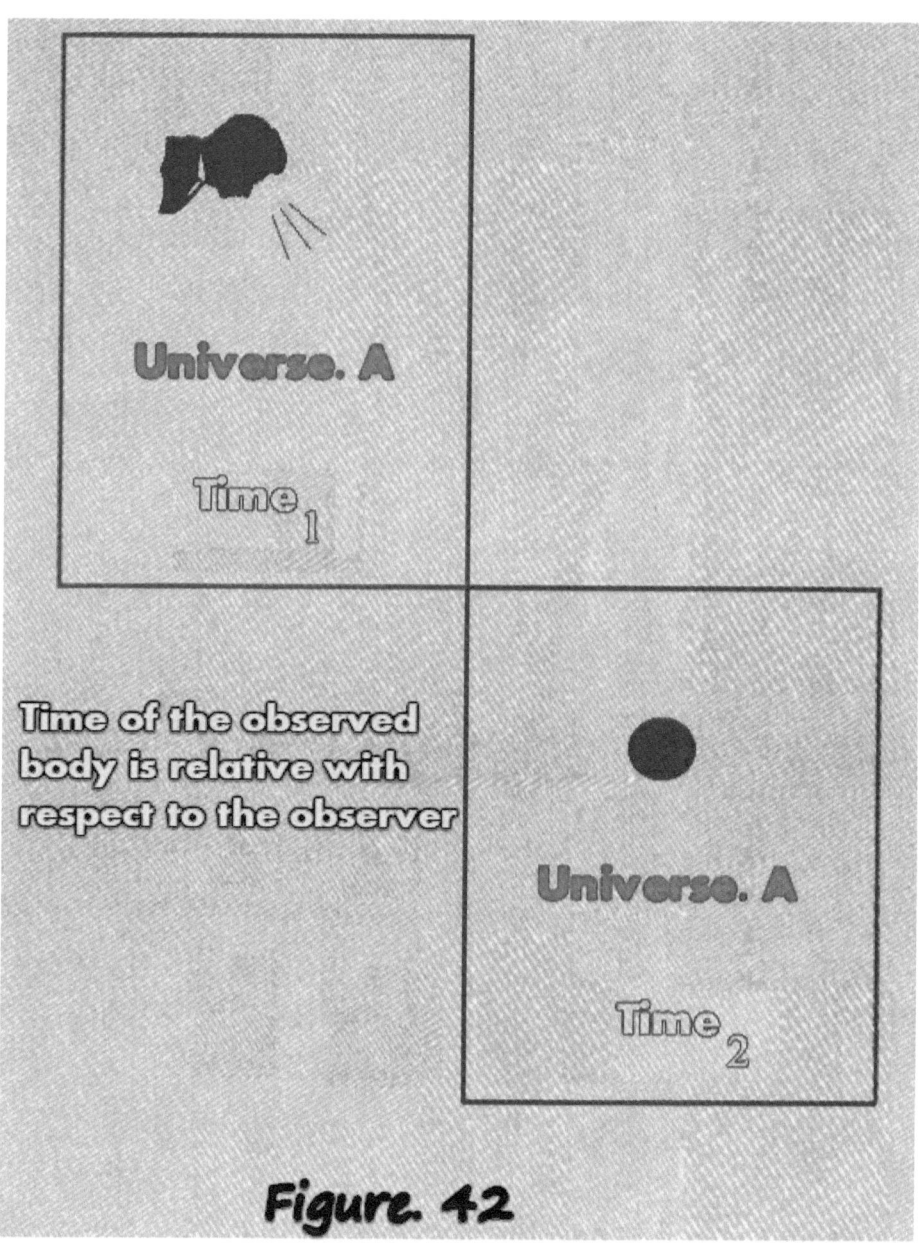

Figure. 42

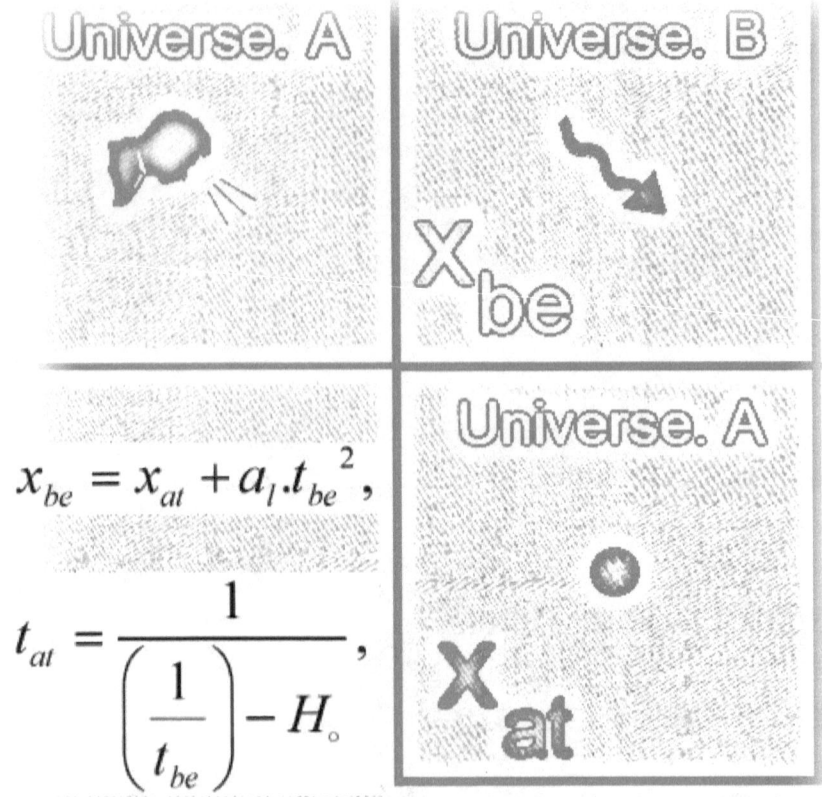

Figure. 43

* * *

CHAPTER 3:

Mathematics of space relativity

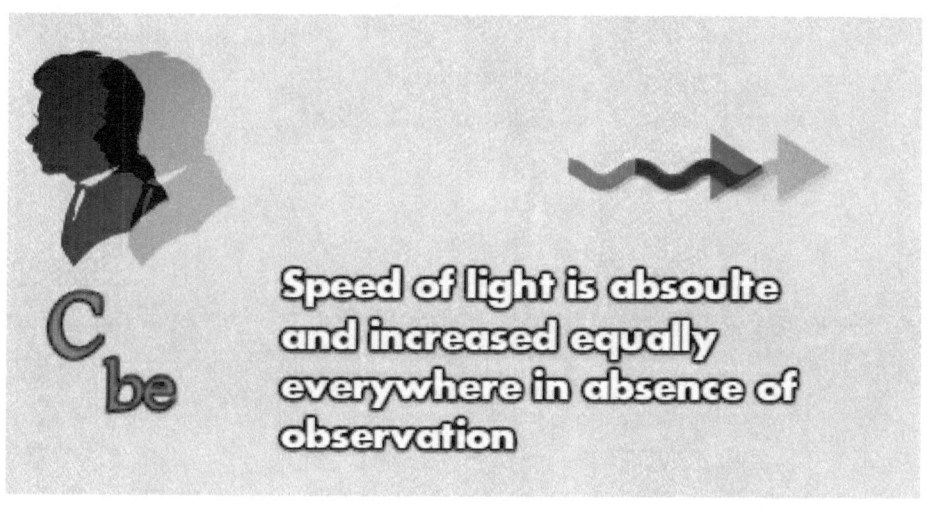

Speed of light is absoulte and increased equally everywhere in absence of observation

Related to the previous physical quantities (σ) in the previous parts or chapters, I find that; it will be more accurate if the word "now" is replaced with "be" and the word "past" is replaced by "at", where (σ_{be}) denotes the physical quantity before obser-vation (or in absence of observation) that equals (σ_{now}), while (σ_{at}) denotes the physical quantity at observation that equals (σ_{past}) (because the observation process takes time and thus it is related to past) as,

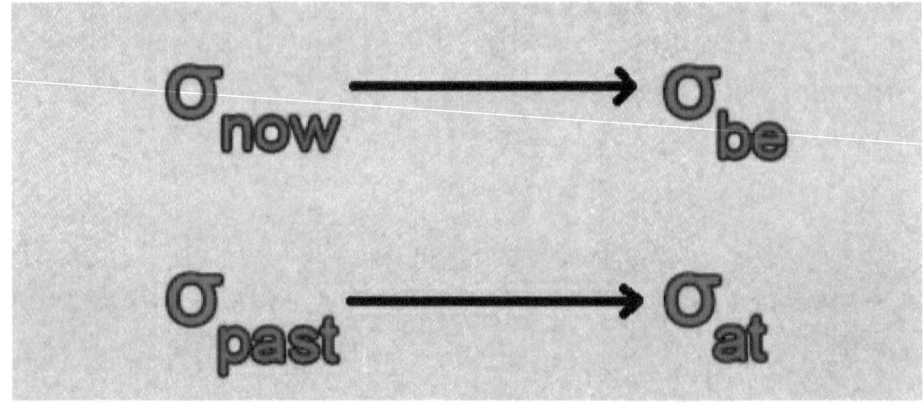

* * *

3.1 Mathematical derivation of the change of position of the particle (Δx)

R egarding to the previous section, all particles share the same time in absence of observation while they are exist in different universe where each universe has spe-cific space (spaces are relative), so if ($x_{universe.A}$) denotes the par-ticle's position in the universe of the observer (at observation) while ($x_{universe.B}$) denotes the particle's position in universe (B) (before observation), the change in particle's position between the two universes (Δx) (uncertainty in position) can be given as,

$$\Delta x = x_{universe.B} - x_{universe.A}, (116)$$

$$\Delta x = c_{universe.B} \cdot t_{be} - c_{universe.A} \cdot t_{be},$$

$$\Delta x = t_{be} \left(c_{universe.B} - c_{universe.A} \right),$$

$$\Delta x = \Delta c \cdot t_{be}, (117)$$

where (t_{be}) denotes the time of the particle before observation or without observation, and (Δc) denotes the change of speed of light.

Note:
Time (t_{be}) is absolute in absence of observation due to the speed of light is absolute in absence of observation.

-(Eq. 117) can be re-written as,

$$\Delta x = \left(a_1 \cdot t_{be} \right) t_{be},$$

$$\Delta x = a_1 \cdot t_{be}^2, (118)$$

✻ ✻ ✻

3.2 Mathematical derivation of wavelength of the particle (λ)

$$z = \frac{\Delta \upsilon}{\upsilon_{at}} , (119)$$

$$z = \frac{1}{\upsilon_{at} . \Delta t} ,$$

And as,

$$z = \frac{\Delta c}{c_{at}} ,$$

Iget,

$$\frac{\Delta c}{c_{at}} = \frac{1}{\upsilon_{at}.\Delta t},$$

$$\Delta c.\Delta t = \frac{c_{at}}{\upsilon_{at}},$$

$$\Delta c.\Delta t = \lambda, (120)$$

Thus, the wavelength of particles is equal to the change of speed of light times the change of time.

* * *

3.3 Relations

3.3.1 The relation between the change of position of the particle (Δx), the change of speed of light (Δc) and acceleration of light (a_1):

$$\Delta x = \Delta c . t_{be,}$$

And as,

$$\Delta c = H_o . D,$$

I get,

$$\Delta x = H_o . D . t_{be,}$$

$$\Delta x = H_o . D . \frac{D}{c_{be}},$$

$$\Delta x = \frac{H_o . D^2}{c_{be}},$$

And as,

$$a_1 = H_o . c_{be},$$

I get,

$$\Delta x = \frac{H_o^2 . D^2}{a_l} \, ,$$

$$\Delta x = \frac{\Delta c^2}{a_l} \, , (121)$$

* * *

3.3.2 The relation between the change of position of the particle (Δx) and its change of momentum (Δp):

$$\lambda_{be} = \Delta c.\Delta t,$$

$$\frac{h}{p_{be}} = \Delta c.\Delta t,$$

$$h = \Delta c.\Delta t.p_{be},$$

$$h = \Delta c\left(t_{be}.z\right)p_{be},$$

$$h = \Delta c.t_{be}\left(p_{be}.z\right),$$

$$h = \left(\Delta c.t_{be}\right)\Delta p,$$

$$h = \Delta x.\Delta p,\left(122\right)$$

Therefore, I find that; if the change of particle's position between two spacetimes is increased, the change of momentum of that particle between these two spacetimes must be decreased regarding to the previous equation.

* * *

3.3.3 The relation between the change of position of the particle (Δx) , the distance (D) and the rate of change of time (z):

$$\Delta x = \Delta c.t_{be},$$

$$\Delta x = \Delta c . \frac{D}{c_{be}},$$

$$\Delta x = D \left(\frac{\Delta c}{c_{be}} \right),$$

$$\Delta x = \frac{D}{1+z} \left(\frac{\Delta c}{c_{at}} \right),$$

$$\Delta x = \frac{D}{1+z} (z),$$

$$\Delta x = D.z \left(1 - H_{\circ} .t_{be} \right), (123)$$

$$z = \frac{\Delta x}{D \left(1 - H_{\circ} .t_{be} \right)}, (124)$$

❊ ❊ ❊

3.3.4 The relation between the change of momentum of the particle (Δp) and the change of speed of light (Δc):

$$\Delta x = \frac{\Delta c^2}{a_1},$$

$$a_l = \frac{\Delta c^2}{\Delta x},$$

And as,

$$h = \Delta x . \Delta p,$$

I get,

$$a_l = \frac{\Delta c^2 . \Delta p}{h},$$

$$a_l . h = \Delta p . \Delta c^2 , (125)$$

Returning to (part. 2, chapter. 4), we find that,

$$a_l . h = Y , (126)$$

Thus (Eq. 125) can be re-written as,

$$Y = \Delta p . \Delta c^2 , (127)$$

where (Y) is a constant that is equal to,

$$Y = \left(6.902138691 \times 10^{-10}\right)\left(6.62607004 \times 10^{-34}\right),$$

$$Y = 4.573405439235991764 \times 10^{-43} m^3 . kg . s^{-3},$$

(Y) can be expressed in another unit that (J.m/s)

Another method to derive the wavelength of particles (λ) using (Eq. 60 & 127)

$$\Delta p.\Delta c^2 = Y = F_5.c_{be}.\lambda_{at},$$

$$\Delta p.\Delta c^2 = F_5.c_{be}.\lambda_{at},$$

$$\Delta p.\Delta c^2 = \frac{\Delta p}{t_{be}}.c_{be}.\lambda_{at},$$

$$\Delta c^2.t_{be} = c_{be}.\lambda_{at},$$

$$\Delta c^2.t_{be} = c_{be}.\frac{\lambda_{be}}{1+z},$$

$$\Delta c^2.t_{be} = c_{at}.\lambda_{be},$$

$$\frac{\Delta c}{c_{at}}.\Delta c.t_{be} = \lambda_{be},$$

$$z.\Delta c.t_{be} = \lambda_{be},$$

$$\Delta c\left(t_{be}.z\right) = \lambda_{be},$$

$$\Delta c.\Delta t = \lambda_{be},$$

✳ ✳ ✳

3.3.5 The relation between the change of position of the particle (Δx) and the fifth force (F₅):

$$h = \Delta x . \Delta p,$$

$$h = \Delta x \left(F_5 . t_{be} \right),$$

$$\frac{h}{t_{be}} = \Delta x . F_5 , (128)$$

where (F5) can be given using (Eq. 58) as,

$$F_5 = m_{be} . a_l \left(1 + z \right),$$

* * *

3.3.6 The relation between the change of speed of light (Δc) and the fifth force (F₅):

$$h = \Delta x . \Delta p,$$

$$h = \left(\Delta c . t_{be} \right)\left(F_5 . t_{be} \right),$$

$$\frac{h}{\left(t_{be} \right)^2} = \Delta c . F_5, (129)$$

$$\frac{h}{\left(t_{be} \right)^3} = a_1 . F_5, (130)$$

❊ ❊ ❊

3.3.7 The relation between the change of position of the particle (Δx) and its wavelength (λ):

$$\lambda_{be} = \Delta c.\Delta t,$$
$$\lambda_{be} = \Delta c \left(t_{be}.z \right),$$
$$\lambda_{be} = \Delta x.z, (131)$$

* * *

3.4 Mathematical derivation of the change of energy of the particle (ΔE)

$$\Delta E = \left(E_{spacetime.B} \right) - \left(E_{spacetime.A} \right),$$

$$\Delta E = \left(p_{be} \cdot c_{spacetime.B} \right) - \left(p_{be} \cdot c_{spacetime.A} \right),$$

$$\Delta E = p_{be} \left(c_{spacetime.B} - c_{spacetime.A} \right),$$

$$\Delta E = p_{be} \cdot \Delta c, (132)$$

* * *

3.5 The relation between the change of energy (ΔE) and the change of time (Δt)

$$\Delta E = p_{be} \cdot \Delta c,$$

And as,

$$p_{be} = \frac{\Delta p}{z},$$

Iget,

$$\Delta E = \frac{\Delta p . \Delta c}{z},$$

And as,

$$h = \Delta x . \Delta p,$$

I get,

$$\Delta E = \frac{h . \Delta c}{\Delta x . z},$$

$$\Delta E \left(\frac{\Delta x . z}{\Delta c} \right) = h,$$

$$\Delta E \left(\frac{\Delta c . t_{be} . z}{\Delta c} \right) = h,$$

$$\Delta E \left(t_{be} . z \right) = h,$$

$$\Delta E . \Delta t = h, (133)$$

* * *

CHAPTER 4:

Mathematics of time relativity

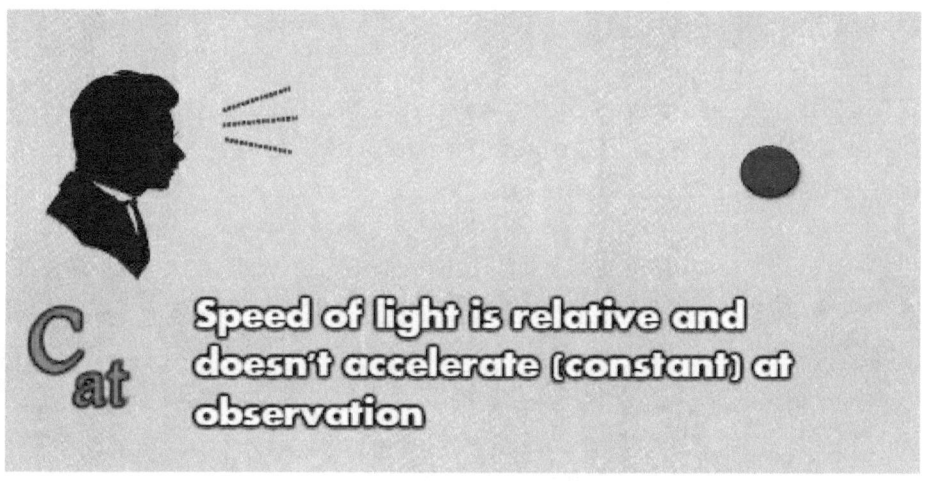

Speed of light is relative and doesn't accelerate (constant) at observation

4.1 Speed of light before observation and at observation

Based on equations in (part. 2, chapter. 1), as the speed of light is increased with time in T.S universe and as the observation process takes time, I find that,

"Speed of light goes to be decreased at observation regarding to the distance that the light travels (or regarding to space between the observer and the observed body)".

If (c_{be}) denotes the speed of light before observation, while (c_{at}) denotes the speed of light at observation, the relation between (c_{be}) and (c_{at}) is given as,

$$c_{at} = \frac{c_{be}}{(1+z)},$$

$$c_{at} = c_{be}\left(1 - H_{o}.t_{be}\right),$$

$$c_{at} = c_{be} - H_{o}.D, (134)$$

where (t_{be}) denotes the time before observation, (Eq. 134) can be expressed in another form as,

$$c_{be} = c_{at}\left(1 + z\right),$$

$$c_{be} = c_{at}\left(1 + H_{o}.t_{at}\right),$$

$$c_{be} = c_{at} + H_{o}.D, (135)$$

where (t_{at}) denotes the time at observation. Thus the transformation of speed of light between "before observation" and "at observation, or in other words, with respect to the observer" can be shown as,

$$c_{at} = c_{be} - H_{\circ}.D,$$

$$c_{be} = c_{at} + H_{\circ}.D, (136)$$

❊ ❊ ❊

4.2 Time of bodies before observation and at observation

B ased on equations in (part. 2, chapter. 2), I find that the contraction of the second with respect to the previous one refers to dilation of time in the past with respect to the present, where the second in the past moves slower with respect to the current second. As the process of observation takes time and as the time moves faster and faster, I find that,

"Time of bodies (or particles) goes to be dilated at observation, or in other words; time of the observed body is dilated with respect to the observer regardless the observed body in motion or stationary".

If (t_{be}) denotes time of the body before observation, while (t_{at}) denotes time of the the body at observation, the relation between (t_{be}) & (t_{at}) is given as,

$$t_{at} = t_{be} \left(1+z\right),$$

$$t_{at} = \frac{t_{be}}{1 - H_o.t_{be}},$$

$$t_{at} = \frac{1}{\left(\dfrac{1}{t_{be}}\right) - H_o}, \quad (137)$$

(Eq. 137) can be expressed in another form as,

$$t_{be} = \frac{t_{at}}{\left(1+z\right)},$$

$$t_{be} = \frac{t_{at}}{1 + H_o.t_{at}},$$

$$t_{be} = \cfrac{1}{\left(\cfrac{1}{t_{at}}\right) - H_{o}}, (138)$$

Thus the transformation of time of the particle between "before observation" and "at observation" can be shown as,

$$t_{at} = \cfrac{1}{\left(\cfrac{1}{t_{be}}\right) - H_{o}},$$

$$t_{be} = \cfrac{1}{\left(\cfrac{1}{t_{at}}\right) - H_{o}}, (139)$$

The general idea in this section that "the further away the observed body from the observer, the slower clock it has with respect to this observer".

4.3 Wavelength of bodies before observation and at observation

B ased on equations in (part. 2, chapter. 1), as the wavelength is increased with time in T.S universe and as the observation process takes time, I find that,

"Wavelength of particles goes to be decreased or contracted at observation, or in other words; wavelength of the observed particle is contracted with respect to the observer".

If (λ_{be}) denotes wavelength of the particle before observation, while (λ_{at}) denotes wavelength of the particle at observation, the relation between (λ_{be}) & (λ_{at}) is given as,

$$\lambda_{at} = \frac{\lambda_{be}}{(1+z)},$$

$$\lambda_{at} = \lambda_{be}\left(1 - H_{\circ}.t_{be}\right), (140)$$

(Eq. 140) can be expressed in another form as,

$$\lambda_{be} = \lambda_{at}\left(1+z\right),$$

$$\lambda_{be} = \lambda_{at}\left(1 + H_{\circ}.t_{at}\right), (141)$$

Thus the transformation of wavelength of the particle between "before observation" and "at observation" can be shown as,

$$\lambda_{at} = \lambda_{be} \left(1 - H_{\circ}.t_{be}\right),$$
$$\lambda_{be} = \lambda_{at} \left(1 + H_{\circ}.t_{at}\right),(142)$$

❊ ❊ ❊

4.4 Momentum of particles before observation and at observation

B ased on equations in (part. 2, chapter. 3), as the momentum of particles is decreased with time in T.S universe and as the observation process takes time, I find that,

"Momentum of particles goes to be increased at observation, or in other words; momentum of the observed particle is increased with respect to the observer".

If (pbe) denotes the momentum of particle before observation, while (pat) denotes the momentum of particle at observation, the relation between (pbe) and (pat) is given as,

$$P_{at} = P_{be}\left(1+z\right),$$

$$P_{at} = \frac{P_{be}}{1 - H_{o}.t_{be}},(143)$$

(Eq. 143) can be expressed in another form as,

$$P_{be} = \frac{P_{at}}{\left(1+z\right)},$$

$$P_{be} = \frac{P_{at}}{1 + H_{o}.t_{at}},(144)$$

Thus the transformation of momentum of the particle between "before observation" and "at observation" can be shown as,

$$P_{at} = \frac{P_{be}}{1 - H_{o}.t_{be}},$$

$$P_{be} = \frac{P_{at}}{1 + H_{o}.t_{at}},(145)$$

❋ ❋ ❋

4.5 Mass of particles before observation and at observation

Based on equations in (part.2, chapter 5), as the mass of the particle is decreased or reduced with time in T.S universe and as the observation process takes time, I find that,

"Particles become heavier at observation, or in other words; mass of the observed particle is increased with respect to the observer".

If (m_{be}) denotes the mass of the particle before observation, while (m_{at}) denotes the mass of the particle at observation, the relation between (m_{be}) and (m_{at}) is given as,

$$m_{at} = m_{be}\left(1+z\right)^{2},$$

$$m_{at} = \frac{m_{be}}{\left(1 - H_{\circ}.t_{be}\right)^2}, (146)$$

(Eq. 146) can be expressed in another form as,

$$m_{be} = \frac{m_{at}}{\left(1 + z\right)^2},$$

$$m_{be} = \frac{m_{at}}{\left(1 + H_{\circ}.t_{at}\right)^2}, (147)$$

Thus the transformation of mass of the particle between "before observation" and "at observation" can be shown as,

$$m_{at} = \frac{m_{be}}{\left(1 - H_{\circ}.t_{be}\right)^2},$$

$$m_{be} = \frac{m_{at}}{\left(1 + H_{\circ}.t_{at}\right)^2}, (148)$$

* * *

4.6 Gravity of particles before observation and at observation

B ased on equations in (part.2, chapter 6), as the gravitational factor is increased with time in T.S universe and as the observation process takes time, I find that,

"Gravitational factor goes to be decreased at observation regarding to the distance that the light travels".

If (Gbe) denotes Gravitational factor before observation while (Gat) denotes Gravitational factor at observation, the relation between (Gbe) and (Gat) is given as,

$$G_{at} = G_{be} - \Delta G,$$

Using (Eq. 79) that is,

$$\Delta G = w.D$$

we obtain,

$$G_{at} = G_{be} - w.D, (149)$$

(Eq. 149) can be expressed in another form as,

$$G_{be} = G_{at} + \Delta G,$$
$$G_{be} = G_{at} + w.D, (150)$$

Thus the transformation of gravitational factor between "before observation" and "at observation" can be shown as,

$$G_{at} = G_{be} - w.D,$$
$$G_{be} = G_{at} + w.D, (151)$$

or,

$$G_{at} = \frac{G_{be}}{(1+z)},$$
$$G_{at} = G_{be}(1 - H_o.t_{be}), (152)$$
$$G_{be} = G_{at}(1+z),$$
$$G_{be} = G_{at}(1 + H_o.t_{at}), (153)$$

＊ ＊ ＊

4.7 Temperature of T.S universe before observation and at observation

B ased on equations in (part.2, chapter 8), as the temperature of T.S universe is decreased with time and as the observation process takes time, I find that,

"Temperature of T.S universe goes to be increased at observation regarding to the distance that the light travels".

If (T_{be}) denotes the temperature of T.S universe before observation, while (T_{at}) denotes the temperature of T.S universe at observation, the relation between (T_{be}) and (T_{at}) is given as,

$$T_{at} = T_{be}\left(1+z\right),$$

$$T_{at} = \frac{T_{be}}{1 - H_\circ.t_{be}}, (154)$$

(Eq. 154) can be expressed in another form as,

$$T_{be} = \frac{T_{at}}{\left(1+z\right)},$$

$$T_{be} = \frac{T_{at}}{1 + H_\circ.t_{at}}, (155)$$

Thus the transformation of temperature of CMB radiation between "before observation" and "at observation" can be shown as,

$$T_{at} = \frac{T_{be}}{1 - H_{\circ}.t_{be}},$$

$$T_{be} = \frac{T_{at}}{1 + H_{\circ}.t_{at}}, (156)$$

* * *

Welcome !